HRT-HOOD™: A Structured Design Method for Hard Real-Time Ada Systems

REAL-TIME SAFETY CRITICAL SYSTEMS

Series Editor:

Hussein Zedan, Department of Mathematics and Computational Sciences, Liverpool John Moores University, Liverpool, U.K.

Vol. 1 Time and Probability in Formal Design of Distributed Systems (H.A. Hansson)

Vol. 2 Towards Verified Systems (J. Bowen, ed.)

Vol. 3 HRT-HOOD™: A Structured Design Method for Hard Real-Time Ada Systems (A. Burns and A. Wellings)

REAL-TIME
SAFETY
CRITICAL
SYSTEMS

HRT-HOOD™: A Structured Design Method for Hard Real-Time Ada Systems

Alan Burns
Andy Wellings
Department of Computer Science
The University of York
Heslington, York, U.K.

™:HOOD is a trademark of the HOOD User Group

1995

ELSEVIER

AMSTERDAM • LAUSANNE • NEW YORK • OXFORD • SHANNON • TOKYO

ELSEVIER SCIENCE B.V.
Sara Burgerhartstraat 25
P.O. Box 211, 1000 AE Amsterdam, The Netherlands

ISBN: 0 444 82164 3

This book is printed on acid-free paper.

Printed in The Netherlands.

Contents

Foreword

The increasing use of computers for real-time control on board spacecrafts has bought with it a greater emphasis on the development methodology used for such systems. By their nature spacecraft control computers have to operate unattended for long periods and because of the programmatics of space, systems are subject to a long development cycle. As a result there are two distinct concerns, the first being that the development approach guarantees functional and timing correctness, the second being that problems, particularly those associated with timing, are considered as early as possible in the spacecraft development life cycle.

The European Space Agency has, for a number of years, encouraged the development of software using HOOD. It was thus a natural next step to investigate the incorporation of time within the existing HOOD framework. This has proven to be very beneficial, and this book describes the approach developed by the authors for handling Hard Real-Time applications. It describes both the background scheduling theory, provides practical examples of its application to real life problems, and demonstrates how it is used in the various phases of the development of Hard-Real Time systems.

Thus I consider that HRT HOOD is a beneficial addition to the armoury of techniques for developing real-time systems, and that this book is a welcome addition to literature in this area which I can recommend for its insight and practical value.

Richard Creasey
European Space Research and Technology Centre

Preface

Background

In January 1991, the European Space Agency commissioned a study into the practicality of using process-based scheduling techniques in an on-board application environment. The Study was undertaken by British Aerospace Space Systems Ltd, The University of York and York Software Engineering Ltd (YSE). It was performed in two consecutive phases. The primary activities of the 1st phase were

1) A review of current scheduling theory, the selection of one theory to be used in the study, and the application of the theory to Ada 83 and Ada 95.
2) The definition and development of a structured design methodology to support the design of hard real-time systems.
3) The definition of software tools to support the scheduling method, and the definition of any changes required to an Ada compiler and its stand-alone run-time support system.

Phase 2 involved:

1) the implementation of the identified tools, and changes to the Ada compiler and run-time system
2) the design and implementation of a realistic case study.

A preliminary version of the structured design method (HRT-HOOD) was produced at the end of Phase 1 (September 1991). During Phase 2 of the project (which was completed in March 1993) much experience was gained with HRT-HOOD and this has caused us to update the method. This book therefore describes HRT-HOOD Version 2.0. The main differences between HRT-HOOD and HRT-HOOD Version 2.0 are:

1) Where HRT-HOOD overlaps with HOOD, we have attempted to update HRT-HOOD to be compatible with HOOD 3.1.
2) HRT-HOOD Version 2.0 addresses issues of distributed systems design.
3) The HRT-HOOD Version 2.0 guidelines on mapping to Ada 83 and Ada 95 have been rewritten to facilitate more efficient code generation. Unlike HOOD 3.1 we still provide detailed mappings.
4) Typographical and minor errors have been corrected.

Book Structure

For any set of application requirements there are potentially many system designs which can satisfy the required properties. A particular solution is usually arrived at by following a design method, either formally or informally. For hard real-time systems the situation is no different. The problem, however, is that current design methods do not adequately address the temporal characteristics of these systems. A hard real-time design method should guide the designer to a solution which can be analysed to ensure that the timing requirements have been met. The goal of this book is to present a structured design method which facilitates the construction and analysis of hard real-time systems. The book consists of three parts and five appendices.

Part 1: Hard Real-Time HOOD

Part 1 summarises our overall approach to the engineering of hard real-time systems, and indicates how the software development life cycle can be modified so that it addresses both functional and non-functional application requirements. Using the object-based framework, we identify the abstractions which must be available to the designer to guide the software development process towards the construction of predictable and analysable systems.

Part 2: Mapping HRT-HOOD to Ada

The overall goal of the project was to show how hard real-time systems can be designed and implemented in Ada. In Part 2 we show how HRT-HOOD designs can be systematically translated into Ada 95 and Ada 83.

Part 3: Case Studies

In this part of the book two case studies are presented. The first involves the control of a pump in a mine drainage system; it is a pedagogical study designed to illustrate many of the features available in HRT-HOOD.

The second is the redesign of a real system so that the method can be evaluated. The system is the attitude and orbital control system for the Olympus Satellite.

Appendices

There are five appendices:

Appendix A summarises the terminology used in this book.

Appendix B gives the rules which ensure that HRT-HOOD designs are consistent.

Appendix C defines the syntax of the Object Description Skeleton in a variant of Backus-Naur-Form.

Appendix D presents a more informal definition of the structure of an Object Description Skeleton.

Appendix E gives a definition of a device driver which is used in the mine drainage case study.

Acknowledgements

The development of HRT-HOOD has benefited from sponsorship from the European Space Agency (Contract Number 9198/90/NL/SF), and the UK Defence Research Agency (Contract Number MOD 2191/023). We gratefully acknowledge the support these projects provided.

The authors would like to thank the following people for their help during the project: Chris Bailey, Eric Fyfe, Paco Gomez Molinero, Tulio Vardanega, Fernando Gonzalez-Barcia, and Pete Cornwell. We would also like to acknowledge Andrew Lister of the University of Queensland who helped lay the foundations for this work.

This book was edited whilst one of the authors (Andy Wellings) was on Research Term at the Department of Computer Science, University of Queensland, Australia. We gratefully acknowledge support of the Department.

Real-time Systems Research at York

Alan Burns and Andy Wellings are members of the Real-Time Systems Research Group in the Department of Computer Science at the University of York (UK). This group undertakes research into all aspects of the design, implementation and analysis of real-time systems. Specifically, the group is addressing: formal and structured methods for development, scheduling theories, reuse, language design, kernel design, communication protocols, distributed and parallel architectures, and program code analysis. The aim of the group is to undertake fundamental research, and to bring into engineering practice modern techniques, methods and tools. Areas of application of our work include space and avionic systems, engine controllers, vehicle control and multi-media systems. Further information about the group's activities can be found via our WWW page:

http://dcpu1.cs.york.ac.uk:6666/real-time/

Many of our reports and papers can also be obtained directly from our ftp site. The address of the site is:

minster.york.ac.uk

(if the name is not recognised then a numeric name can be used: `144.32.128.41`)

The reports and papers are located in the directory:

/pub/realtime/papers

Part 1
Hard Real-Time HOOD

The objective of this Part of the book is to illustrate how structured design methods can be tailored towards the construction of real-time systems in general, and hard real-time systems in particular. We use the term hard real-time systems to mean those systems which have components which must produce timely services; failure to produce a service within the required time interval may result in severe damage to the system or the environment, and may potentially cause loss of life (for example, in avionics systems). Rather than developing a new method from scratch, the HOOD (Hierarchical Object Oriented Design) method is used as a baseline. The new method, called HRT-HOOD (Hard Real-Time HOOD), was designed as part of an European Space Agency (ESA) supported project. HOOD was chosen as the base-line because ESA currently recommend the use of HOOD for their systems development. However, we believe the ideas presented in the book can be used to extend other common design methods such as Mascot.[38]

1 Overview of the HRT-HOOD Design Process

1.1. Introduction

The most important stage in the development of any real-time system is the generation of a consistent design that satisfies an authoritative specification of requirements. In this, real-time systems are no different from other computer applications. However, real-time systems do differ from the traditional data processing systems in that they are constrained by certain non-functional requirements (e.g. dependability and timing). Typically the standard design methods do not have adequate provisions for expressing these types of constraints.

There are a number of ways of classifying forms of design notation (or representation). For our purposes McDermid[66] gives a useful decomposition. He names three techniques:

- informal
- structured
- formal

Informal methods usually make use of natural language and various forms of imprecise diagrams. They have the advantage that the notation is understood by a large group of people (i.e. all those speaking the natural language). It is well known, however, that phrases in natural language text, for example, are often open to a number of different interpretations.

Structured methods often use a graphical representation, but unlike the informal diagrams these graphs are well-defined. They are constructed from a small number of predefined components which are interconnected in a controlled manner. The graphical form may also have a syntactical representation in some well defined language. HOOD,[4] JSD,[53] and MASCOT[52,75] are examples of structured design methods often used for real-time systems.

Although structured methods can be made quite rigorous they cannot, in themselves, easily be analysed or manipulated. It is necessary for the notation to have a mathematical basis if such operations are to be carried out. Methods that have such mathematical properties are usually known as formal. They have the clear advantage that precise descriptions can be made in these notations. Moreover, it is possible to prove that necessary properties hold; for example, that the top-level design satisfies the requirement specification. The disadvantage with formal techniques is that they can not handle complex systems at present. Furthermore, they cannot easily be understood by those not prepared or able to become familiar with the notation.

The high reliability requirements in real-time systems has caused a movement away from informal approaches to the structured and, increasingly, the formal. Rigorous techniques of verification and validation are beginning to be used in the real-time industry but, at present, few software engineers have the necessary mathematical skills to exploit fully their potential.

There are many structured design methods which are targeted toward real-time systems: MASCOT, JSD, Yourdon, MOON, HOOD, DARTS, MCSE etc. None of these, however, support directly the common hard real-time abstractions (such as periodic and sporadic activities) which are found in most hard real-time systems. Neither do they impose a computational model that will ensure that effective timing analysis of the final system can be undertaken. Consequently, their use is error prone and can lead to systems whose real-time properties cannot be analysed. For a comparison of these methods see Hull et al,[41] Cooling,[37] and Calvez.[34] PAMELA (Process Abstraction Method for Embedded Large Application)[36] is perhaps the exception. It allows for the diagrammatic representation of cyclic activities, state machines and interrupt handlers. However, the notation is not supported by abstractions for resources and therefore designs are not necessarily amenable to schedulability analysis.

There are a few CASE environments available to support the real-time systems design process; Teamwork,[80] for example, provides an environment in which designs can be expressed using Buhr's graphical notation[17] or HOOD. Tools which analyse the timing properties of the resulting designs, however, are not included. EPOS[60] is another environment which supports the whole real-time system life cycle. It provides three specification languages: one for describing the customer requirements, one for describing the system specification, and one for describing project management, configuration management and quality assurance. Like HRT-HOOD, the system specification has a graphical and a textual representation. Both periodic and sporadic activities can be represented (but they are not directly visible when manipulating the diagrams) and timing requirements can be specified. Furthermore, early recognition of the real-time behaviour of applications has been addressed within the context of EPOS;[61] however, this work relies on animating the system specification rather than schedulability analysis.

One project which has addressed some of the issues raised by this book is MARS.[38] Kopetz et al[54] present a design method that is influenced by a real-time transaction model. A transaction is triggered by an event (usually occurring in the environment of the system) and results in a corresponding response. During design, a transaction is refined into a sequence of subtransactions, finally becoming a set of tasks which may be allocated to processing units (called "clusters"). Attributes such as deadline, criticality etc can be associated with a transaction. A system life cycle is proposed which carries out both dependability and schedulability analysis.

1.2. The Importance of Non-Functional Requirements

It is increasingly recognised that the role and importance of non-functional requirements in the development of complex critical applications has hitherto been inadequately appreciated. Specifically, it has been common practice for system developers, and the methods they use, to concentrate primarily on functionality and to consider non-functional requirements comparatively late in the development process. Experience shows that this approach fails to produce safety critical systems. For example, often timing requirements are viewed simply in terms of the performance of the completed system. Failure to meet the required performance often results in ad hoc changes to the system. This is not a cost effective process.

Non-functional requirements include dependability[59] (e.g. reliability, availability, safety and security), timeliness (e.g. responsiveness, orderliness, freshness, temporal predictability and temporal controllability), and dynamic change management[57] (i.e. incorporating evolutionary changes into a non-stop system). These requirements, and the constraints imposed by the execution environment, need to be taken into account throughout the system development life cycle. During development an early binding of software function to hardware component is required so that the analysis of timing and reliability properties of a still unrefined design can be carried out.[55]

We believe that if hard real-time systems are to be engineered to high levels of dependability, a real-time design method must provide:

- the explicit recognition of the types of activities/objects that are found in hard real-time systems (i.e. cyclic and sporadic activities);
- the integration of appropriate scheduling paradigms with the design process;
- the explicit definition of the application timing requirements for each object;
- the explicit definition of the application reliability requirements for each object;
- the definition of the relative importance (criticality) of each object to the successful functioning of the application;
- the support for different modes of operation — many systems have different modes of operation (e.g., take-off, cruising, and landing for an aircraft); all the timing and importance characteristics will therefore need to be specified on a per mode basis;
- the explicit definition and use of resource control objects;
- the decomposition to a software architecture that is amenable to processor allocation, schedulability and timing analysis;
- facilities and tools to allow the schedulability analysis to influence the design as early as possible in the overall design process;
- tools to restrict the use of the implementation language so that worst case execution time analysis can be carried out;
- tools to perform the worst case execution time and schedulability analysis.

1.3. The Software Development Life Cycle

In the this section we show how the above key aspects can be addressed by extending the usual software development life cycle. Most traditional software development methods (including HOOD) incorporate a life cycle model in which the following activities are recognised:

- Requirements Definition — during which an authoritative specification of the system's required functional and non-functional behaviour is produced. It is beyond the scope of this book to address issues of requirements capture and analysis. We assume that the techniques for identifying objects and their operations are those described in the HOOD User Manual.[2]
- Architectural Design — during which a top-level description of the proposed system is developed.
- Detailed Design — during which the complete system design is specified.
- Coding — during which the system is implemented.
- Testing — during which the efficacy of the system is tested.

For hard real-time systems, this has the significant disadvantage that timing problems will only be recognised during testing, or worse after deployment.

A constructive way of describing the process of system design is as a progression of increasingly specific *commitments*.[18,42] These commitments define properties of the system design which designers operating at a more detailed level are not at liberty to change. Those aspects of a design to which no commitment is made at some particular level in the design hierarchy are effectively the subject of *obligations* that lower levels of design must address. Early in design there may already be commitments to the structure of a system, in terms of object definitions and relationships. However, the detailed behaviour of the defined objects remains the subject of obligations which must be met during further design and implementation.

The process of refining a design — transforming obligations into commitments — is often subject to *constraints* imposed primarily by the execution environment (Figure 1.1). The execution environment is the set of hardware and software components (e.g. processors, task dispatchers, device drivers) on top of which the system is built. It may impose both resource constraints (e.g. processor speed, communication bandwidth) and constraints of mechanism (e.g. interrupt priorities, task dispatching, data locking). To the extent that the execution environment is immutable these constraints are fixed.

Obligations, commitments and constraints have an important influence on the architectural design of any application. We therefore define two activities of the architectural design:

- the logical architecture design activity;
- the physical architecture design activity.

The logical architecture embodies commitments which can be made independently of the constraints imposed by the execution environment, and is primarily aimed at satisfying the functional requirements (although the existence of timing requirements, such as end-to-end deadlines, will strongly influence the decomposition of the logical architecture).

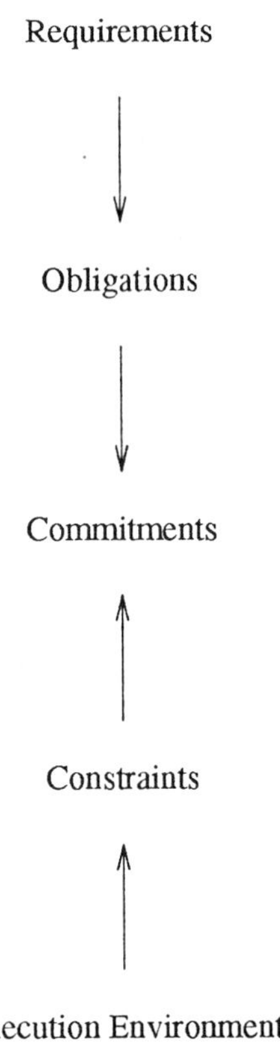

Figure 1.1: Obligations, Commitments and Constraints

The physical architecture takes these functional requirements and other constraints into account, and embraces the non-functional requirements. The physical architecture forms the basis for asserting that the application's non-functional requirements will be met once the detailed design and implementation have taken place. It addresses timing and dependability requirements, and the necessary schedulability analysis that will ensure (guarantee) that the system once built will function correctly in both the value and time domains (within some failure hypotheses). To undertake this analysis, it will be necessary to make some initial estimations of the execution time of the proposed code (and other resource requirements such as LAN usage), and to have available the time dependent behaviour of the target processor and other aspects of the execution environment. Dependability analysis evaluates the design with respect to reliability, safety and security.

Although the physical architecture is a refinement of the logical architecture its development will usually be an iterative and concurrent process in which both models are developed/modified. The analysis techniques embodied in the physical architecture can, and should, be applied as early as possible. Initial resource budgets can be defined that are then subject to modification and revision as the logical architecture is refined. In this way a 'feasible' design is tracked from requirements through to deployment.

Once the architectural design activities are complete, the detailed design can begin in earnest and the code for the application produced. When this has been achieved, the execution profile of the code must again be estimated (using a worst case execution time analyser tool) to ensure that the initial estimated worst case execution times are indeed accurate. If they are not (which will usually be the case for a new application), then either the detailed design must be revisited (if there are small deviations), or the designer must return to the architectural design activities (if serious problems exist)†. If the estimation indicates that all is well, then testing of the application proceeds. This should involve measuring the actual time of the code's execution. The modified life cycle is presented in Figure 1.2.

Detailed Design and Coding should follow the usual process, although code measurement for worst case timing behaviour is a complex issue. It will be necessary to constrain the way in which code is written so that analysis of execution time of the final system can be carried out in a way that is not too pessimistic. For example all loops must be bounded, and there must be limited recursion only.

1.4. Summary

The chapter has defined a design process that is applicable to hard real-time systems. We have distinguished between two phases in the architectural design activity — the logical phase and the physical phase.

The logical architecture phase addressed the functional activities of the system. The physical architecture addresses the non-functional (e.g. timing) requirements and the constraints of the execution environment. This includes the hardware (eg speed and memory capacity) and the functionality of the kernel if this is a fixed entity. The activities undertaken during the physical architecture phase will define priorities, offsets and (timing) error conditions that will be accommodated. It may also lead to the functionality of some objects being altered, and the inclusion of new objects. The result is that timing behaviour is guaranteed.

HRT-HOOD attempts to be independent of any scheduling theory that might be used to guarantee the timing properties of programs (although in the current project we are using preemptive *priority*-based scheduling analysis). Instead, it provides a framework within which the properties of real time applications can be expressed. Similarly it does not prescribe the approach by which fault tolerance is achieved (other than by the provision of exception handlers and the possible replication of objects).

Once the architectural design phase is complete detail design and coding will ensue. The key issue here is to constrain the coding style to one that is amenable to (non-pessimistic) worst case execution time analysis.

---

† Of course as the detailed design progresses, it may be necessary to revisit the logical and physical architectures to provide new or modified functionality (possibly as a result of a change in requirements).

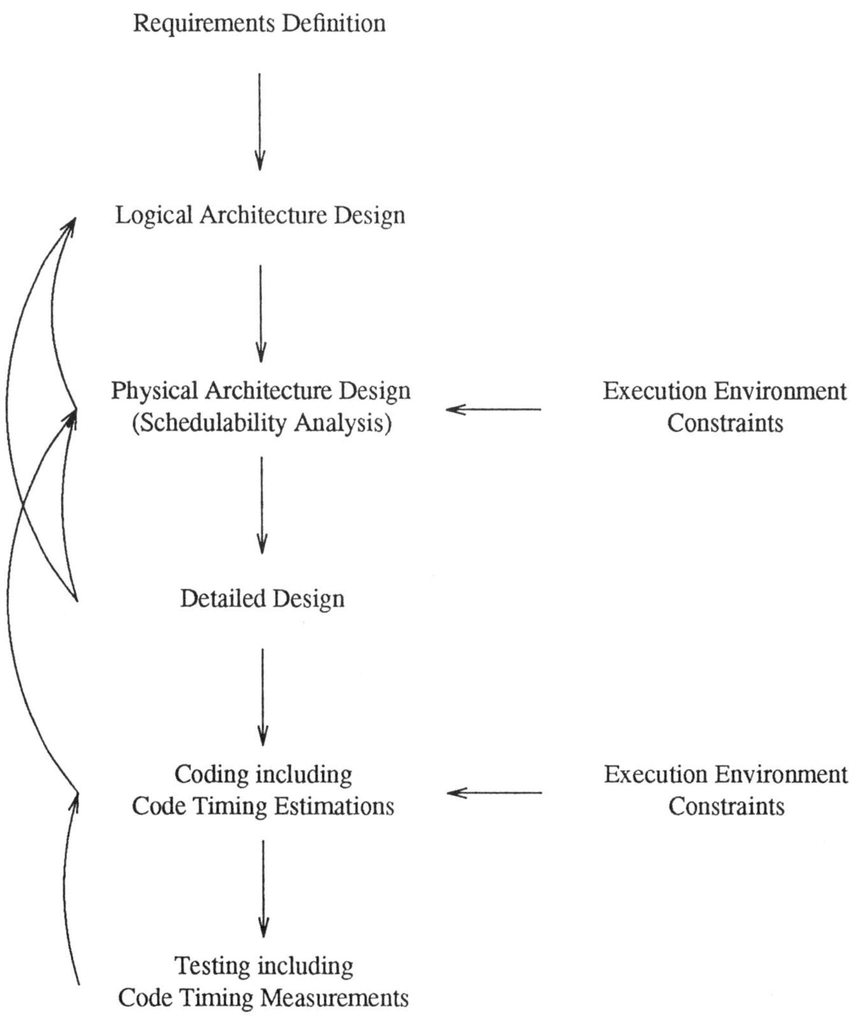

Figure 1.2: The Hard Real-Time Life Cycle

2 Logical and Physical Architecture Design in HRT-HOOD

In this chapter we consider the support required for logical and physical architecture design.

2.1. Logical Architecture Design

There are two aspects of any design method which facilitate the logical architecture design of hard real-time systems. Firstly, explicit support must be given to the abstractions that are typically required by hard real-time system designers. *We take the view that if designs are to be well structured so that they can be analysed, then it is better to provide specific design guidelines rather then general design abstractions.* For example, supporting the abstraction of a periodic activity allows the structure of that activity to be visible to the design process which, in turn, facilitates its analysis. In contrast, allowing the designer to construct periodic activities out of some more primitive "task" activity produces designs which are more difficult to analyse. Clearly, care must be taken to ensure that the design method does not become cluttered with too many abstractions, however, it is important that the design method allows the expression of important analysable program structures.

The second aspect involves constraining the logical architecture so that it can be analysed during the physical architecture design activity. In particular designs are forced to map down to a computational model which facilitates analysis. For example, HRT-HOOD supports single threaded objects which interact via data-oriented communication and synchronisation mechanisms (rather than via tightly synchronous communication).

These aspects are now discussed in detail.

2.1.1. Supporting Common Hard Real-Time Abstractions

The outcome of the logical architecture design activity is a collection of terminal objects (objects which do not require further decomposition) with all their interactions fully defined. It is assumed that some form of functional decomposition process has led to the definition of these terminal objects. Although this decomposition is essentially functional, the existence of timing requirements is a major driver in the construction of the application's architecture. Timing requirements are usually derived from other (higher-level) requirements such as controllability. They normally take the form of deadlines between input and output actions. These input/output relationships also represent functional behaviour, and hence the initial decomposition of the system can be along functional lines.

In HOOD only two basic object types are defined: *passive* and *active*. HRT-HOOD extends the number of base types to include: *protected*, *cyclic* and *sporadic*. They are defined as follows.

- *Passive* — objects which have no control over when invocations of their operations are executed, and do not spontaneously invoke operations in other objects.
- *Active* — objects which may control when invocations of their operations are executed, and may spontaneously invoke operations in other objects. *Active* objects are the most general class of objects and have no restrictions placed on them.
- *Protected* — objects which may control when invocations of their operations are executed, and do not spontaneously invoke operations in other objects; in general *protected* objects may *not* have arbitrary synchronisation constraints and must be analysable for the blocking times they impose on their callers.
- *Cyclic* — objects which represent periodic activities, they may spontaneously invoke operations in other objects, but the only operations they have are requests which demand immediate attention (they represent asynchronous transfer of control, ATC, requests).
- *Sporadic* — objects which represent sporadic activities; *sporadic* objects may spontaneously invoke operations in other objects; each sporadic has a *single* operation which is called to invoke the sporadic, and one or more operations which are requests which demand immediate attention (they represent asynchronous transfer of control requests).

HRT-HOOD distinguishes between the synchronisation required to execute the operations of an object and any internal independent concurrent activity within the object. The synchronisation agent of an object is called the Object Control Structure (OBCS). The concurrent activity within the object is called the object's *thread*. The thread executes independently of the operations, but when it executes operations the order of the executions is controlled by the OBCS.

A hard real-time program designed using HRT-HOOD will contain at the terminal level only *cyclic*, *sporadic*, *protected* and *passive* objects. *Active* objects, because they cannot be fully analysed, will only be allowed for background activity. *Active* object types may also be used during decomposition of the main system but must be transformed into one of the above types before reaching the terminal level.

Cyclic and *sporadic* activities are common in real-time systems; each contains a single *thread* that is scheduled at run-time. *Protected* objects control access to data that is shared by more than one thread (i.e. *cyclic* or *sporadic* object); in particular they provide mutual exclusion. *Protected* objects are similar to monitors[48] and conditional critical regions[16] in that they can block a caller if the conditions are not correct for it to continue. A *passive* object can be used concurrently if this can be achieved within out error. Alternatively it may be used by just one object or by a collections of object — but never at the same time.

With these types of terminal objects other common paradigms used in hard real-time systems can be supported, in particular precedence constrained activities. These involve a series of computations through terminal objects. They are likely to occur in a design which must reflect *transaction* deadlines. For example, consider a reactive system which on receipt of a value from its input sensors must produce an output value to an actuator within a certain period. The production of the actuator setting might require the coordination of more than one object (often in the form of precedence constraints). The term transaction is used here to represent the totality of computation and communication required to produce the output.

At a high level of design, a transaction is represented by a single object: *cyclic* or *sporadic* depending on whether the transaction is time driven or event driven. The object is then decomposed into a set of terminal precedence constrained objects. If the first activity of the transaction is cyclic then there is a *cyclic precedence constrained activity* (or *cyclic transaction*); if the first activity were sporadic (e.g., event triggered) then it would be a *sporadic precedence constrained activity* (or *sporadic transaction*). The key property of precedence constrained activities is that any activity within it can start immediately its predecessor has terminated. Hence, there is a collection of *before* and *after* relationships.

In summary, the logical architecture design process may commence with the production of *active* and *passive* objects, and by a process of decomposition will lead to the production of terminal objects of the appropriate character. Transactions are therefore represented as *cyclic* or *sporadic* objects which decompose to terminal objects with precedence constrained and related activities.

2.1.2. Constraining the Design for Analysis

In order to analyse the full design certain constraints are required. These are mainly concerned with the allowed communication/synchronisation between objects. They are:

(a) *cyclic* and *sporadic* objects may not call arbitrary blocking operations in other *cyclic* or *sporadic* objects.

(b) *cyclic* and *sporadic* objects may call operations which effect an asynchronous transfer of control operations in other *cyclic* or *sporadic* objects.

(c) *protected* objects may not call blocking operations in any other object.

(d) *passive* operations contain only sequential code which does not need to synchronise with any other object.

Where appropriate the design method itself prohibits the above behaviours.

Asynchronous transfer of control is used to get the *immediate* attention of a thread. The details of how it is implemented is, however, a concern for the implementation language and the underlying operating system (or kernel). It can be used during fast mode changes but is not a general requirement; many applications will not make use of the facility.

Other constraints may be placed on the design to aid dependability analysis. For example, to facilitate object replication it may be necessary to ensure that a particular object is deterministic.

It is important to emphasise that any hard real-time design method must constrain the design process if it is to produce analysable software. HRT-HOOD is explicitly designed to ensure that system decomposition conforms to a set of constraints that facilitates analysis of the final system.

2.2. Physical Architecture Design

The physical architectural design activity is concerned with mapping the logical architecture onto the required physical resources in the target system†. In order to undertake the necessary analysis, we must:

- have a design expressed in a manner that facilitates analysis
- have a means of predicting the behaviour of the design on a given platform (hardware and kernel).

The logical architecture design activity ensures that the design conforms to a computational model which facilitates timing analysis. The exact form this analysis must take is not defined by HRT-HOOD. We have, however, successfully integrated the HRT-HOOD design process with the use of static priority analysis and preemptive dispatching.[6, 27, 28] The construction of cyclic or best-effort scheduling would also be possible with the computational model. In general physical architecture design is concerned with four activities:

1) object allocation — the allocation of objects in the logical architecture to processors within the constraints imposed by the functional and non-functional requirements (e.g, ensuring that all device controller objects are located on the sites where the controlled devices reside, or that all replicas run at separate sites)
2) network scheduling — scheduling the communications network so message delays are bounded
3) processor scheduling — determining the schedule (static or dynamic) which will ensure that all tasks within all objects residing on all processors will meet their deadlines
4) dependability — for example, determining whether an object should be replicated to tolerate hardware failures, estimating the complexity of an object to see whether software fault tolerant techniques should be employed

In general, the three activities of task allocation, processor scheduling and network scheduling are all NP-hard problems.[23] This has led to a view that they should be considered separately. Unfortunately, it is often not possible to obtain optimal solutions (or even feasible solutions) if the three activities are treated in isolation. For example, allocating a task T to a processor P will increase the computational load on P, but may reduce the load on the communications media (if T communicates with tasks on P), and hence the response time of the communications media is reduced, allowing communications deadlines elsewhere in the system to be met. The tradeoffs can become

† For some designs the emphasis of the physical architectural design activity may be on determining the minimal resources required to meet all the timing and dependability requirements. The ideas presented in this section equally apply to this case.

very complex as the hard real time architecture becomes more expressive. It is beyond the scope of this book to discuss these issues. The reader is referred to the relevant literature.[13,32,35,39,69,85]

Whatever allocation and scheduling method is used, the design method must support the definition of a physical architecture by:

1) allowing timing attributes to be associated with objects,

2) providing the abstractions with which the designer can express the handling of timing (and other run-time) errors

The physical design must of course be feasible within the context of the execution environment. This is guaranteed by the allocation and schedulability analysis.

Issues of dependability must also be addressed during this activity. Both hardware and software failures must be considered. As an example of the issues addressed in the construction of the physical architecture, suppose that the reliability requirements together with knowledge of the hardware's failure characteristics imply that a particular object be replicated (so that a copy is always available, given the system failure hypotheses). The replicated object is a single entity in the logical architecture; in the physical architecture it becomes several objects, each mapped to a particular processor. Suppose further that the chosen approach to obtaining the required level of reliability is such that the replication is 'active', that is, all copies must always have the same state. The physical architecture expresses this property, and makes explicit the obligations:

a) that the object is deterministic, and

b) that an atomic broadcast protocol must be provided or implemented.

Other forms of replication such as leader/follower[14] or passive replication can also be specified. Each will present specific obligations.

2.2.1. Object Attributes

The way in which the non-functional requirements are specified during the physical architecture design activity is via object attributes. All terminal objects have associated real-time attributes. Many attributes are associated with mapping the timing requirements on to the logical design (e.g., deadline, importance). These must be set before the schedulability and dependability analysis can be performed. Other attributes (such as priority, required replication etc) can only be set during this analysis.

For each specified mode of operation, the *cyclic* and *sporadic* objects have a number of temporal attributes defined:

- The period of execution for each *cyclic* object.
- The minimum arrival interval for each *sporadic* object.
- Offset times for related objects.
- Deadlines for all sporadic and cyclic activities.

Two forms of deadline are identified. One is applied directly to a sporadic or cyclic activity. The other is applied to a precedence constrained activity (transaction); here there is a deadline on the whole activity and hence only the last activity has a true deadline. The deadlines for the other activities must be derived so that the complete transaction satisfies its timing requirements (in all cases).

To undertake the schedulability analysis, the worst case execution time for each thread and all operations (in all objects) must be known. After the logical design activity these can be estimated (taking into account the execution environment constraints) and appropriate attributes assigned. Clearly, the better the estimates the more accurate the schedulability analysis. Good estimates come from component reuse or from arguments of comparison (with existing components on other projects). During detailed design and coding, and through the direct use of measurement during testing, better estimates will become available which typically will require the schedulability analysis to be redone.

In general not all activities within the system will be at the same level of criticality. *Cyclic* and *sporadic* objects will therefore be annotated with a criticality level (e.g. safety critical, mission critical, background). As well as providing valuable information for the schedulability analysis, the criticality level may also be used during the validation and verification of the total design. Critical activities may receive more rigorous analysis of both their timing characteristics and their functional characteristics, than less critical activities. Note that although *passive* and *protected* objects do not have a directly associated criticality (importance) level, for validation and verification purposes, they will be assigned a criticality level that equals the highest of their users.

During the physical architecture design activity it will be necessary to commit to a run-time scheduling approach, for example static priority scheduling with priorities assigned using rate or deadline monotonic scheduling theory.[6,7,47,64] This requires other attributes to be supported. For example the priority of *cyclic* and *sporadic* objects. *Protected* objects may be implemented by having ceiling priorities assigned that are at least as high as the maximum priority of the *threads* that use the operations defined on that object.[68]

In general the physical architecture design is concerned with annotating (via attributes) the objects contained in the logical design. There are, however, a number of cases in which extra objects may be added during this activity. In addition to replication for availability, it may be desirable to reduce output jitter (a non-functional requirement); a *cyclic* object (say) with period T and deadline D, which could produce its output anywhere between its minimal execution time and D, may be replaced by two related objects and a *protected* object. The first object which contains all the functionality of the original will have a deadline of D1 (D1 < D) by which time it will have placed the output data in a *protected* object. The second *cyclic* process, which actually performs the output and which also has a cyclic time of T, will have an offset of D1 and a deadline of D-D1. The closer D1 is made to D the smaller the jitter.

There is a final important point to emphasise about the activity of architectural design that we have called the physical design. Although there are benefits from seeing it as a distinct activity, it will typically proceed concurrently with the logical architectural design. It may be natural to add timing requirements to objects as they are defined. However, the full analysis can only be undertaken once a complete set of terminal objects is known. Furthermore, the overheads associated with the proposed execution environment must be taken into account when estimating execution times etc.

2.2.2. Handling Timing (and other run-time) Errors

Schedulability analysis can only be effective if the estimations/measurements of worst case execution time is accurate. Within the timing domain two strategies can be identified for limiting the effects of a fault in a software component:

- Do not allow an object to use more computation time (budget time) than it requested.
- Do not allow an object to execute beyond its deadline.

One would expect a design method to allow a designer to specify the actions to be taken if objects overrun their allocated execution time or miss their deadlines. In both of these cases the object could be informed (via an exception) that a timing fault will occur so that it can respond to the error (within the original time frame, be it budget or deadline). There is an obligation on the execution environment to undertake the necessary time measurements and to support a means of informing an object that a fault has occurred. There is also an obligation on the coding language to provide primitives that will allow recovery to be programmed.

Sporadic objects present further difficulties if they execute more often than was envisaged when the schedulability analysis was carried out. There is an obligation on the execution environment to ensure that a *sporadic* object does not execute too early and to allow recovery action to be taken if invocations come too frequently.

A common system paradigm that makes use of *sporadic* objects is interrupt driven I/O. The logical architecture will map the interrupt to the non-blocking start operation of a *sporadic* object. The arrival of an interrupt will therefore release the *sporadic* object. For the schedulability analysis, undertaken as part of defining the physical architecture, the interrupt will be modeled as a sporadic action with a minimum arrival interval. Interrupts can also be mapped to operations on *protected* objects, which may further release *sporadic* objects.

If the failure hypothesis of the hardware system is that interrupts will not occur too often then the above model is adequate. However if the software system wishes to protect itself from a fault that would be caused by an over-active interrupt source then it must disable interrupts for a defined period.

2.2.3. Other Forms of Analysis

The emphasis so far in this chapter has been on the timing requirements of hard real-time systems. There will, however, be other forms of analysis that should be carried out as part of the definition of a physical architecture (even for a single processor system). If there is a fixed memory constraint then this will impose obligations on the detailed design and coding. Where power consumption is an issue then the amount of RAM must be limited, and the constructive use of ROM may need to be addressed. Even weight constraints may have an influence on the physical architecture (this would be particularly true if distribution or multiprocessors were an option). We believe, however, that the framework introduced in this section can be extended into these areas.

2.3. Summary

The logical architecture phase supports hierarchical decomposition (or refinement) and provides the definition of a collection of *cyclic*, *sporadic*, *protected* and *passive* terminal objects. *Cyclic* and *sporadic* objects will each contain a single thread that is dispatched at run-time. An important aspect of the logical architecture is that it constrains the interactions between *threads so that schedulability analysis can be carried out in the next phase.*

The physical architecture addresses the non-functional (e.g. timing) requirements and the constraints of the execution environment. The activities undertaken during the physical architecture phase will define priorities, offsets and (timing) error conditions that will be accommodated. It may also lead to the functionality of some objects being altered, and the inclusion of new objects. The result is that timing behaviour is guaranteed.

HRT-HOOD attempts to be independent of any scheduling theory that might be used to guarantee the timing properties of programs (although in the material presented in this book we are using preemptive *priority* based scheduling analysis). Instead, it provides a framework within which the properties of real time applications can be expressed. Similarly it does not prescribe the approach by which fault tolerance is achieved (other than by the provision of exception handlers and the possible replication of objects).

3 HRT-HOOD Objects

In this section HRT-HOOD objects, their graphical representation, their *real-time attributes* and their relationships are described in more detail. The rules defining a valid HRT-HOOD design are given in Appendix A.

3.1. Graphical Representation

Figure 3.1 illustrates the overall graphical representation of an HRT-HOOD object (this is almost identical to that of HOOD, the only difference being the addition of the object type indicator). The textual representation of an object is defined by the *Object Description Skeleton* (ODS) which is described in Appendices C and D.

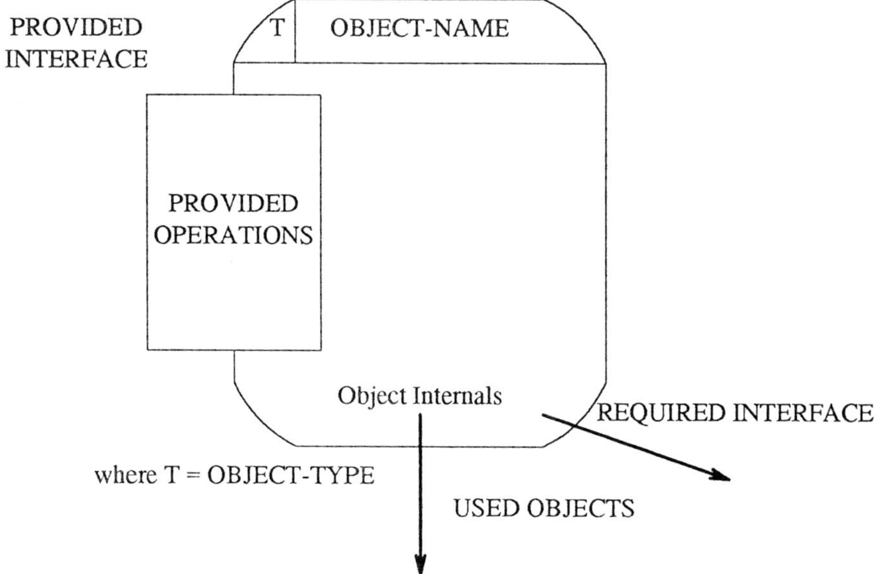

Figure 3.1: The Graphical Representation of an HRT-HOOD Object

The interface to an object has two components: the *provided interface* and the *required interface*. The *provided interface* defines the types, constants, operations and exceptions provided by the object, and the associated parameters and types. The *required interface* indicates the other objects, types, constants, operations and exceptions required by the object.

The internals of an object consist of the implementation of the *provided operations* which were described in the interface. They may use *internal operations* and data (if they are *terminal objects*), as well as the types, constants, operations, and exceptions required from other objects. In HRT-HOOD, as in HOOD, the internals of an operation is called its *OPeration Control Structure* (OPCS).

HRT-HOOD uses the term *use* to indicate that one object requires another in order to implement its internals; this is graphically illustrated in Figure 3.2. The *use* relation is discussed further in Section 3.8, and Appendix A, Section A.1 defines the rules associated with its use.

In order to provide top-down decomposition of a system, a parent object is decomposed into a set of child objects that collectively provide the same functionality as the parent. The *root* object represents the system to be designed. An object which has no children is called a *terminal* object. Intermediate objects are termed *non-terminal* objects. The *Design Process Tree* is the tree of the system being designed, and consists of the *root object* and its successive decomposition into child objects until *terminal objects* are reached.

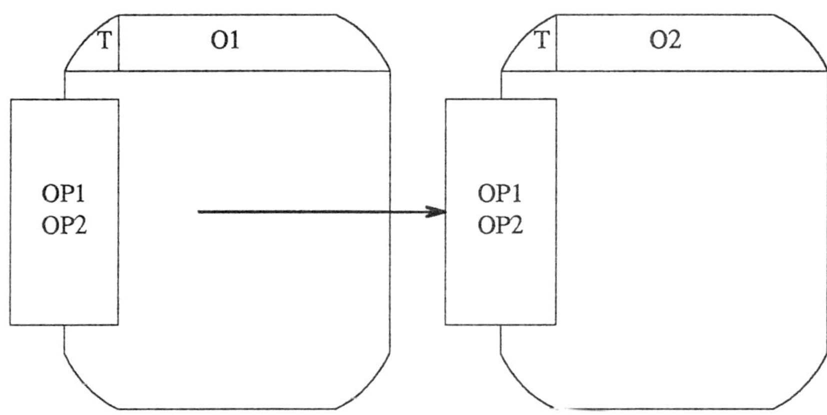

Figure 3.2: The Use Relationship

The decomposition process is based on the *include* relationship, a parent *includes* a child. The graphical representation of the *include* relationship is as follows:

- the included relationship is represented by drawing the child objects inside the parent object
- the mapping between the operations at the parent level and the child level is shown by a shaded or dashed arrow
- there is no graphical representation for the OBCS (the synchronisation agent) or the *thread*.

The *include* relationship is discussed further in Section 3.9, and Appendix A, Section A.3 define the rules governing the *include* facility.

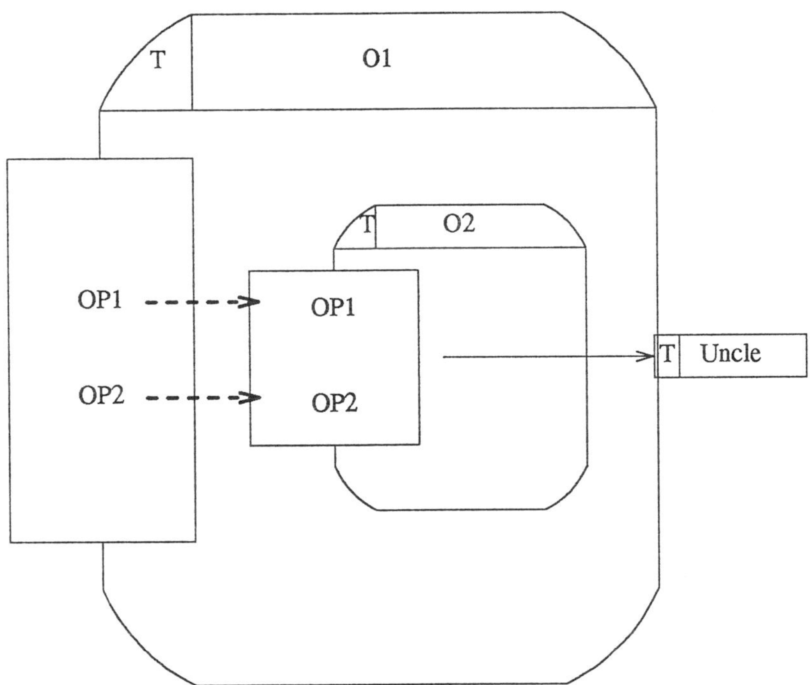

Figure 3.3: The Include Relationship and Uncles

Figure 3.3. shows an example of a parent and child object. Note that a child object may use an object of a higher level of decomposition (but only if the parent uses the object). Such an object is usually referred to as an *uncle* object and it also graphically illustrated in Figure 3.3.

3.2. Passive Objects

Passive objects have no control over when their operations are executed. That is, whenever an operation on a passive object is invoked, control is immediately transferred to that operation. Each operation contains only sequential code which does not synchronise with any other object (i.e. it does not block). *A passive object has no OBCS and no thread.* (Figure 3.4 illustrates the graphical representation of a *passive* object.)

3.3. Active Objects

The operations on an *active* object may be constrained or unconstrained. *Unconstrained operations* are executed as soon as they are requested, i.e. they are similar to the operations on a *passive* object.

Constrained operations are executed under the control of the OBCS and the *thread*. As with HOOD, there are two classes of constraint that can affect when operations are executed.

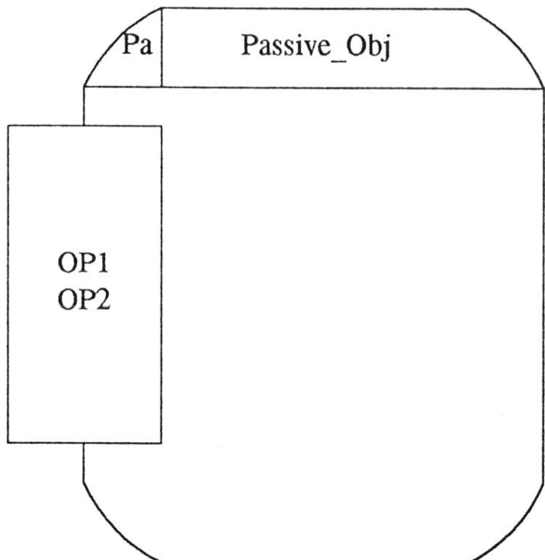

Figure 3.4: The Graphical Representation of a *Passive* Object

1) *Functional activation constraints* impose constraints on when an invoked operation can be executed according the object's internal state. An operation is said to be "open" if the object's internal state allows the operation's execution (for example: a buffer object would allow a "put" operation if its internal storage was not full). An operation is said to be "closed" if the object's internal state does not allow the operation's execution. An operation which has no functional activation constraints is considered to be "open".

   Operations which have *functional activation constraints* are marked by an asterisk in the HRT-HOOD diagrams.

   A calling object is blocked if the called operation is closed and the operation has an LSER or HSER type of request constraint (see below).

2) *Type of request* constraints indicate the effect on the caller of requesting an operation. The following request types are supported:

   Asynchronous Execution Request — ASER

   When an ASER operation is called, the caller is not blocked by the request. The request is simply noted and the caller returns. In a process/message-based design method this would be equivalent to asynchronous message passing.

   Loosely Synchronous Execution Request — LSER

   When an LSER operation is called, the caller is blocked by the request until the called object is ready to service the request. In a process/message-based design method this would be equivalent to the occam/CSP synchronous style of message passing.

   Highly Synchronous Execution Request — HSER

When an HSER operation is called, the caller is blocked by the request until the called object has serviced the request. In a process/message-based design method this would be equivalent to the Ada extended rendezvous style of message passing (or remote invocation[19]).

Both LSER and HSER may have an associated timeout in which case they are termed TOER_LSER (*Timed Operation Execution Request* LSER) or TOER_HSER.

Operations which have *type of request* constraints have an associated trigger arrow and a type of request label shown in the HRT-HOOD diagram.

In general a *constrained operation* may have a *functional activation constraint*, and will always have a type of request constraint.

Figure 3.5 gives the graphical representation of an *active* object, where OP1 and OP2 are *constrained operations* and OP3 is unconstrained. OP1 also has functional activation constraints.

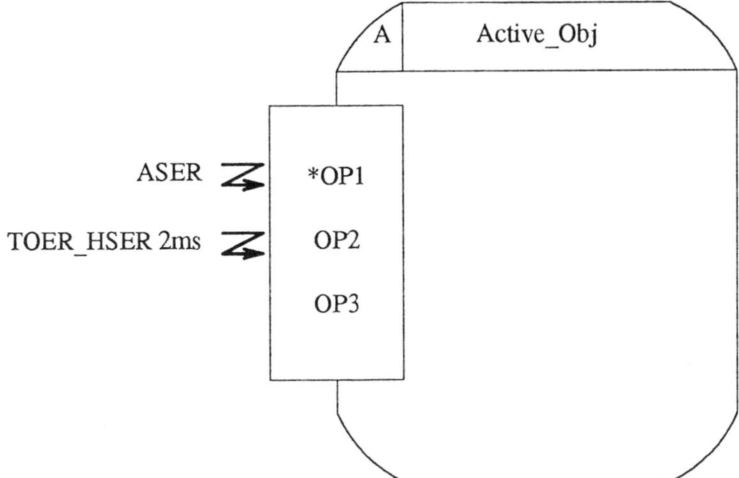

Figure 3.5: The Graphical Representation of an *Active* Object

3.4. Protected Objects

Protected objects are used to control access to resources which are used by hard real-time objects. The intention is that their use should constrain the design so that the run-time blocking for resources can be bounded (for example by using *priority* inheritance,[72] or some other limited blocking protocol such as the immediate priority ceiling inheritance associated with the Ada 95 protected types[51]).

Protected objects are objects which are able to control when their invoked operations are executed, but (unlike *active* objects) they do not necessarily require independent threads of control. A *protected* object does have an OBCS but this is a monitor-like construct: operations may require mutual exclusion, and functional activation constraints may be placed on when operations can be invoked. For example, a bounded buffer might be implemented as a *protected* object.

Two types of *constrained operation* are available on *protected* objects. They are:

- *Protected synchronous execution request* (PSER).
- *Protected asynchronous execution request* (PAER).

A *constrained operation* at some point in its execution will require mutually exclusive access to the object's data. PAER operations, because they are asynchronous, can only have "in" parameters; PSER can support the full range of parameter types.

Only a PSER type of request can have a functional activation constraint which can impose any required synchronisation. A functional activation constraint which evaluates to true is said to be "open", one which evaluates to "false" is said to be closed. A timeout PSER request (TOER_PSER) must have a functional activation constraint.

It should be noted that in general having arbitrary synchronisation constraints on protected objects will lead to designs that cannot be analysed for their timing characteristics. If functional activation constraints are used for operations of terminal hard real-time protected objects then these operations will usually be declared with a timeout. However, where protected objects are being used to implement mode changes, cyclic or sporadic objects may be blocked without a timeout.

Protected objects may also have non-*constrained operations*, which are executed in the same manner as *passive* operations.

Figure 3.6 gives the graphical representation of a *protected* object.

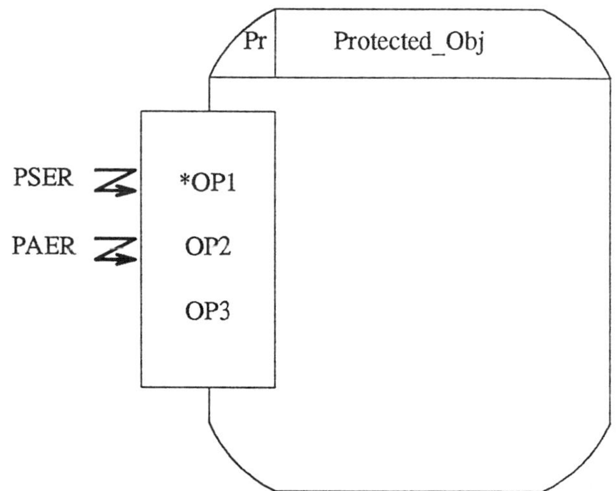

Figure 3.6: The Graphical Representation of a *Protected* Object

3.5. Cyclic Objects

Cyclic objects are used to represent periodic activities. They are *active* objects in the sense that they have their own independent threads of control. However, these *threads* (once started) execute irrespective of whether there are any outstanding requests for their objects' operations. Furthermore, they do not wait for any of their objects' operations at any time during their execution. Indeed, in many cases *cyclic* objects will not have any operations.

In general the *thread* of a *cyclic* objects will communicate and synchronise with other hard real-time *threads* by calling operations in *protected* objects. However, it is recognised that some *constrained operations* may be defined by a *cyclic* object to allow other objects to communicate directly with the *cyclic* object. In particular:

- other objects may need to signal a mode change to the cyclic object — this could be achieved by having *cyclic* objects poll a "mode change notifier" *protected* object but this is inefficient if the response time required from the *cyclic* object is short (if mode changes can occur only at well defined instances then "mode change notifier" objects would be appropriate)
- other objects may need to signal error conditions to the cyclic object — this could be achieved by having *cyclic* objects poll an error notifier *protected* object but this is again inefficient when the response time required from the *cyclic* object is short.

All *constrained operations* declared by *cyclic* objects require an urgent response from the *cyclic*'s *thread*. The OBCS of a *cyclic* object interacts with the *thread* to force an asynchronous transfer of control. Although these "operation execution requests" require urgent attention, it is possible that there will be some delay before the object responds (for example, the object may be executing a critical region and does not want to be disturbed). Consequently we identify several types of *constrained operations* (each may have functional activation constraints). These directly correspond to the types identified for *active* objects, and they indicate the effect on the caller of issuing the request. Available operations are†:

- *Asynchronous, asynchronous transfer of control request* (ASATC) — The caller issues the request and continues with its execution immediately.
- *Loosely synchronised asynchronous transfer of control request* (LSATC) — The caller issues the request and waits until the called object has acknowledged the request and is ready to act upon it.
- *Timed-out loosely synchronised asynchronous transfer of control request* (TOER_LSATC) — TOER_LSATC operation allows the caller to time-out waiting for the request to be acknowledged.
- *Highly synchronised asynchronous transfer of control request* (HSATC) — The caller issues the request and waits for the request to be satisfied.
- *A Timed-out highly synchronised asynchronous transfer of control request* (TOER_HSATC) — A TOER_HSATC operation allows the caller to time-out waiting for the request to be satisfied (see TOER_LSATC above).

Note, that if an object is not subject to asynchronous transfer of control request then it will not have an interface.

---

† It is not clear whether the method should support all these operation types. In practice, only ASATC may be required.

It is possible for one of the above operations to occur whilst the *cyclic thread* is blocked awaiting its next *period* of execution. In this case the *thread* is *not* rescheduled until its next *period*. To reschedule immediately would invalidate the schedulability analysis.

A *cyclic* object may start its execution immediately it is created, it may have an *offset* (that is a time before which the *thread* should be delayed before starting its cyclic execution), or it may synchronise its start via a *protected* object (using an PSER operation with a *functional activation constraint*). All *cyclic* objects have a *thread*, whereas only those with operations have an OBCS.

Figure 3.7 graphically represents a *cyclic* object.

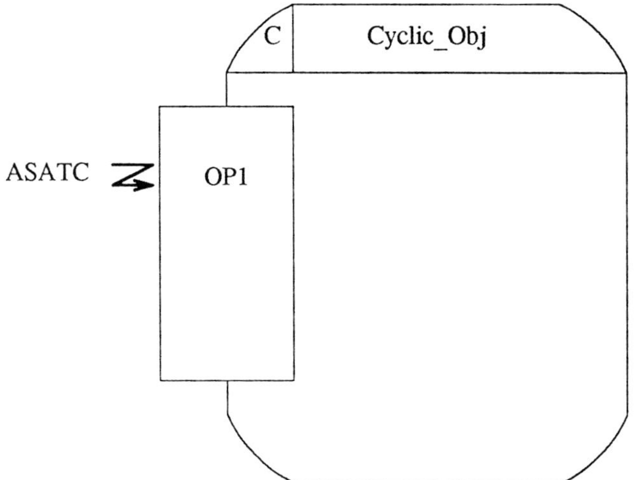

Figure 3.7: The Graphical Representation of a *Cyclic* Object

3.6. Sporadic Objects

Sporadic objects are *active* objects in the sense that they have their own independent threads of control. Each *sporadic* object has a single constrained operation, usually called START, which is called to invoke the execution of the *thread*. The operation is of a type which does not block the caller (ASER); it may be called by an interrupt, in which case the label becomes ASER_BY_IT. The operation which invokes the sporadic has a defined minimum arrival interval, and/or a maximum arrival rate.

A *sporadic* object may have other *constrained operations* but these are requests which wish to affect the *sporadic* immediately to indicate a result of a mode change or an error condition. As with *cyclic* objects ASATC, LSATC, TOER_LSATC, HSATC, and TOER_HSATC operations are all possible. A *sporadic* object which receives an asynchronous transfer of control request will "immediately" abandon its current computation, if executing and not within a protected object.

For the operation which invokes the *sporadic*, there is either a defined minimum arrival interval, or a maximum arrival rate. If a *sporadic* object receives an asynchronous transfer of control request whilst it is in its enforced delay, then the request is not handled until the delay expires (to do otherwise would violate its minimum inter arrival interval).

Sporadic objects may also have non-*constrained operations*, which are executed in the same manner as *passive* operations.

Figure 3.8 graphically represents a *sporadic* object.

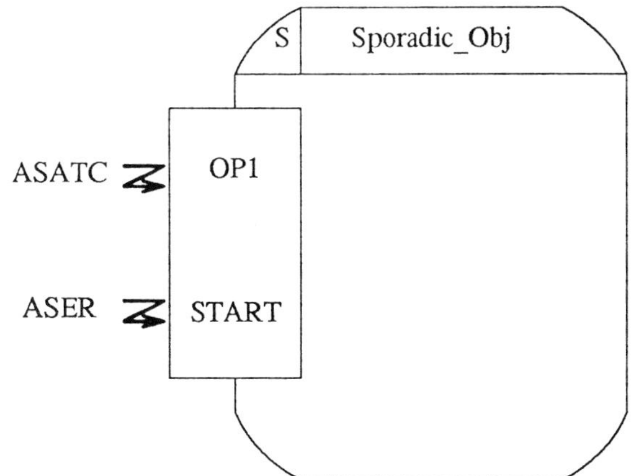

Figure 3.8: The Graphical Representation of a *Sporadic* Object

3.7. Real-Time Object Attributes

HOOD does not explicitly support the expression of many of the constraints necessary to engineer real-time systems. In the object description language there is a field in which the designer can express "implementation and synchronisation" constraints. Rather than use this to express an object's *real-time attributes*, a separate REAL-TIME ATTRIBUTES field has been added. These attributes are filled in by the designer normally at the *terminal object* level. It is anticipated that many of the values of the attributes will be computed by support tools.

The following attributes are required (note, the system, in general, can operate in several distinct modes; thus a number of attributes are given on a per mode basis):

- DEADLINE

  Each *cyclic* and *sporadic* object may have a defined *deadline* for the execution of its *thread*. If the object can exist in several modes, there may be a list of *deadlines* (one for each mode).

- OPERATION BUDGET

  Each externally visible operation of an object may have a budget execution time defined. If the object can exist in several modes, and this is transparent at the object's interface, then there may be a list of budget execution times for each operation (one for each mode).

  An operation which overruns its budgeted time is terminated. Each externally visible operation of an object, therefore, may have an *internal operation* which is to be called if the operation's budget execution time is violated.

- OPERATION WCET

  Each externally visible operation of an object may have a worst case execution time defined. The *worst case execution time* for the external operation is the operation's *budget time* plus the *budget time* of any internal error handling operation. If the object can exist in several modes, and this is transparent at the object's interface, then there may be a list of *worst case execution times* for each operation (one for each mode). An operation which overruns its WCET is terminated.

- THREAD BUDGET

  Each *cyclic* and *sporadic* object may have a budget execution time defined for each activation of its *thread* of execution. If the object can exist in several modes, there may be a list of budget execution times (one for each mode). An overrun of the budgeted time results in the termination of the activity being undertaken.

  Each *cyclic* and *sporadic* object may have an *internal operation* which is to be called if its *thread*'s budget execution time is violated.

- THREAD WCET

  Each *cyclic* and *sporadic* object may have a worst case execution time defined for its *thread* of execution. The *worst case execution time* for the *thread* is the *thread*'s *budget time* plus the *budget time* of the internal error handling operation. If the object can exist in several modes, there may be a list of *thread worst case execution times* (one for each mode). If the *thread* overruns its WCET then it is terminated for the current invocation.

- PERIOD

  Each *cyclic* object must have a defined *period* of execution. If the object can exist in several modes, there must be a list of *periods* (one for each mode).

- OFFSET

  Each *cyclic* object may have a defined *offset* which indicate the time that the *thread* should delay before starting its cyclic operations. Each *sporadic* object may also have a defined *offset* which indicate the time that the *thread* should delay before starting each invocation.

- MINIMUM_ARRIVAL_TIME or MAXIMUM_ARRIVAL_ FREQUENCY

  Each *sporadic* object must have either a defined minimum arrival time for requests for its execution, and/or a maximum arrival frequency of request. If the object can exist in several modes, there must be a list of times (one for each mode).

- PRECEDENCE CONSTRAINTS

  A *thread* may have *precedence constraints* associated with its execution. This attribute indicates which object must execute before and after it.

- PRIORITY

  Each *cyclic* and *sporadic* object must have a defined *priority* for its *thread*. This *priority* is defined according to the scheduling theory being used (for example, according to the *thread*'s *period* or its *deadline*). If the object can exist in more than one mode then there must be a list of priorities (one for each mode).

- CEILING PRIORITY

  Each *protected*, *cyclic* or *sporadic* object must have a defined *ceiling priority*. This *priority* is no lower than the maximum *priority* of all the *threads* that can call the object's constrained operations. If the *protected*, *cyclic* or *sporadic* object can exist in more than one mode then the *ceiling priority* must be no lower than the maximum *priority* of all the *threads* that can call the constrained operations in all possible modes of operation.

- EXECUTION TRANSFORMATION

  A *cyclic* or a *sporadic* object may need to be transformed at run-time to incorporate extra delays. This may be required, for example, as a result of *period* transformation during the schedulability analysis phase of the method. If the object can exist in more than one mode then there must be a list of transformations (one for each mode).

- IMPORTANCE

  Each *cyclic*, *sporadic* and *protected* object must have a defined *importance*. This *importance* represents whether the object is a hard real-time *thread* or a soft real-time object. If the object can exist in more than one mode then there must be a list of *importance* (one for each mode).

- INTEGRITY

  Each object may have an assigned integrity level. These levels may be defined according to any Industrial Standard being used to support the development process. For example, UK Defence Standard 00-56[40] indicates four integrity levels for software (S1, S2, S3, and S4) and gives rules for how the safety integrity of a high level function shall be inherited by components which implement it. These rules can be enforced by the HRT-HOOD include relationship.

This list may be extended - for example some HRT approaches may require minimum/average execution times, utility functions etc.

3.8. The Use Relationship (Control Flow)

As with HOOD, *passive* objects must not use *constrained operations* of other objects, and *active* objects may use operations of any other object freely. Furthermore, HRT-HOOD forbids *passive* (or *protected*) objects to use each other in a cyclic manner.

The use relationships for the other object types are as follows.

- *Cyclic* and *sporadic* objects must not use *constrained operations* of *terminal active* objects unless they are asynchronous; they can, however, use *constrained operations* of *non-terminal active* objects if these operations are implemented by child *protected* objects. *Cyclic* and *sporadic* objects may use *constrained operations* of *protected* objects

  In any hard real-time system the blocking time of a *thread* must be bounded. Consequently operation calls which may result in arbitrary blocking must be prohibited. In the proposed method, hard real-time *threads* communicate and synchronise with each other via the *constrained operations* on *protected* objects. The method defines a maximum blocking time for each constrained operation). *Active* objects are not considered by the method to be real-time, and consequently *cyclic* and *sporadic* objects must not be dependent upon their execution.

- *Cyclic* and *sporadic* objects may call the *constrained operations* of other *cyclic* or *sporadic* objects.

  These operations may result in some blocking. The worst case response time of an asynchronous transfer of control request to each *cyclic* object should be definable. Timeouts may be used to bound this time if the worst case response is unacceptable.

- *Protected* objects may only use *unconstrained operations*, ASER *constrained operations*, or other *protected* object operations.

  In general *protected* objects should not block once they are executing (although some requeuing of requests may be allowed).

The following table summarises the HRT-HOOD use relationship. Note that any object may use an *unconstrained operation* of any other object.

caller/used	*cyclic*	*sporadic*	*protected*	*active*	*passive*
cyclic	allowed	allowed	allowed	ASER operations only	allowed
sporadic	allowed	allowed	allowed	ASER operations only	allowed
protected	ASATC operations only	ASATC operations only or START	allowed	ASER operations only	allowed
active	allowed	allowed	allowed	allowed	allowed
passive	not allowed	not allowed	not allowed	not allowed	allowed

Table 3.1: Allowed Use Relationship

3.9. The Include Relationship (Decomposition)

The "include relationship" for HRT-HOOD objects are as follows:

- An *active* object may include any other object.
- A *passive* object may only include other *passive* objects.†

† Note that in HOOD a *passive* object may include an *active* object as long as the passive nature of the parent is not violated. HRT-HOOD removes this feature as it is no longer necessary, given HRT-HOOD's increase in expressive power.

- A *protected* object may include *passive* objects, and one *protected* object.

  The intention is that *protected* objects should not have separate *threads* of control. Consequently they can not decompose into objects which have their own independent *threads*. Furthermore, a parent *protected* object guarantees that its operations have mutually exclusive access to its protected data; any decomposition must not violate the parent's guarantee.

- A *sporadic* object may include at least one *sporadic* object along with one or more *cyclic*, *passive*, and *protected* objects.

  At one level of the design an object may be considered *sporadic*. However, it may be decomposed into a group of precedenced constrained *sporadic* objects (communicating via *protected* objects) as long as the *sporadic* nature of the parent is not violated.

- A *cyclic* object may include at least one *cyclic* object along with one or more *passive*, *protected* and *sporadic* objects.

  At one level of the design an object may be considered *cyclic*. However, it may be decomposed into a group of *precedenced constrained cyclic*, *passive*, *protected* and *sporadic* objects as long as the *cyclic* nature of the parent is not violated.

The HRT-HOOD include rules are summarised in the table below.

Parent/Children	*cyclic*	*sporadic*	*protected*	*active*	*passive*
cyclic	allowed	allowed	allowed	not allowed	allowed
sporadic	allowed	allowed	allowed	not allowed	allowed
protected	not allowed	not allowed	allowed	not allowed	allowed
active	allowed	allowed	allowed	allowed	allowed
passive	not allowed	not allowed	not allowed	not allowed	allowed

Table 3.2: Allowed Include Relationship

3.10. Operation Decomposition

3.10.1. Internal Operations

For *terminal* objects the OPCS contains the description of the pseudo-code related to the operation. The OPCS may use *internal operations*; these are operations which may be used in the implementation of an object's interface (the *provided* operations) but are themselves not visible in the interface.

Internal operations are not visible in the HRT-HOOD diagrams but are declared in the OPERATION field of the ODS for *terminal* objects. An *internal operation* of a *protected* operation may indicate if it requires mutual exclusion over the object's data.

Each *terminal* object may have a parameterless "initialisation" operation which gets called when the object is elaborated.

3.10.2. General Scheme

One parent operation is *implemented by* one child operation. The graphical representation is a dashed (or shaded) arrow from the parent operation to the child operation. In the ODS the OPCS of the parent operation has the key words *implemented_by* and the name of the child operations which implement it.

Note that in HOOD, a parent operation can be implemented by one or more child operations by the use of an *op_control* object. This is not allowed in HRT-HOOD as HRT-HOOD ensures that blocking is limited. An *op_control* objects could be introduced but it would need to be of a particular synchronisation type (e.g. PSER, ASER) and the child operation it would be allowed to call would have to be full defined. In effect the *op_control* object becomes a full blown object. Consequently, they are disallowed (i.e. if they are needed they can be implemented as full objects).

A *constrained operation* shall be implemented by a constrained child operation. The following decompositions are valid:

Decomposition	Comments
ASER → ASER ASER → ASATC ASER → PSER	a parent asynchronous operation can be implemented by a child asynchronous operation, or an asynchronous asynchronous transfer of control operation, or or a child synchronous protected operation(see note 1 below)
ASER → PAER	a parent asynchronous can map to a child asynchronous protected operation
*ASER → *ASER *ASER → PSER *ASER → PAER *ASER → *ASATC	an asynchronous request must never block
LSER → LSER LSER → PSER LSER → *PSER LSER → LSATC	a parent loosely synchronous operation can be implemented by a child loosely synchronous operation, a protected synchronous request, a protected synchronous request with a *functional activation constraint*, or a loosely synchronous asynchronous transfer of control operation
*LSER → *LSER *LSER → *PSER *LSER → *LSATC	similar to the above
HSER → HSER HSER → PSER HSER → *PSER HSER → HSATC	a parent highly synchronous operation can be implemented by a child highly synchronous operation, a protected synchronous request, a protected synchronous with a *functional activation constraint*, or a highly synchronous asynchronous transfer of control operation
*HSER → *HSER	similar to the above

Decomposition	Comments
*HSER → *PSER *HSER → *HSATC	similar to the above
PSER → PSER *PSER → *PSER PAER → PAER	*constrained operations* on a parent *protected* object must be implemented by identical operations in the child (see note 2 below)

Table 3.3: Valid Operation Decomposition

Note 1: This is allowed as on a single/multi processor the time taken to execute the operation is small. For a distributed system this decomposition would force the two objects onto the same processor or multiprocessor.

Note 2: All the *constrained operations* of a parent *protected* object must eventually call the same child *protected* object, in order to guarantee the mutual exclusive access to the parent's protected data.

Figures 3.9. - 3.12 show some examples of object and operation decomposition.

- Figure 3.9 shows the decomposition of an *active* object into Iactive *and Ipassive* objects.
- Figure 3.10 shows the decomposition of an *active* object into *cyclic*, *protected* and *passive* objects.
- Figure 3.11 shows the decomposition of an *active* object into *cyclic*, *sporadic* and *passive* objects.
- Figure 3.12 shows the decomposition of a *cyclic* into *cyclic*, *protected* and *passive* objects.

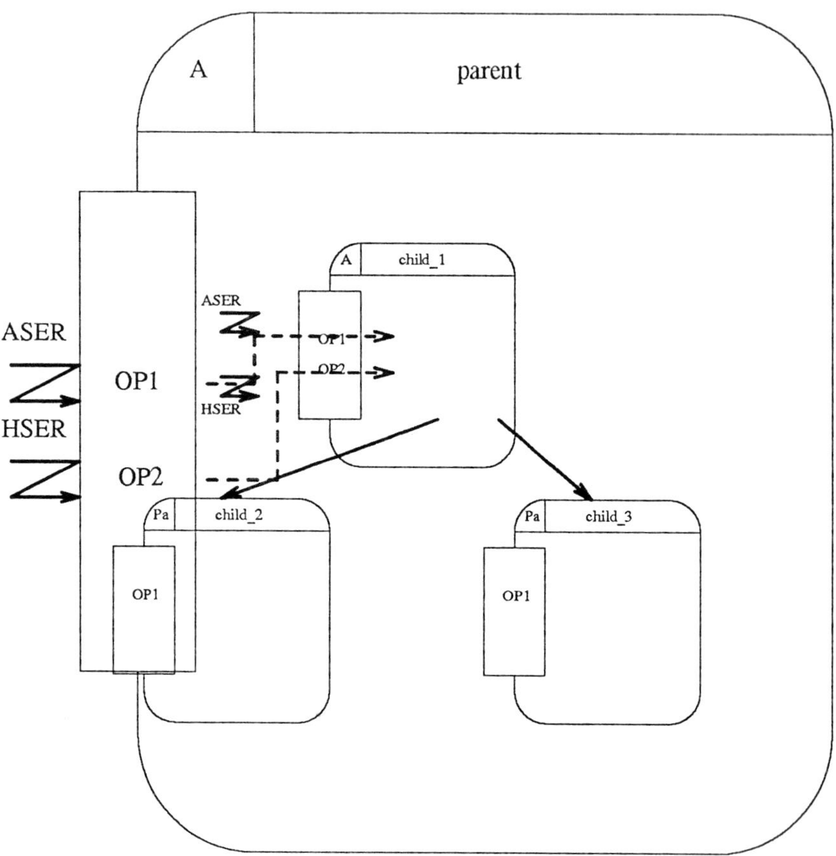

Figure 3.9: The Decomposition of an *Active* Object into *Active* and *Passive* Objects

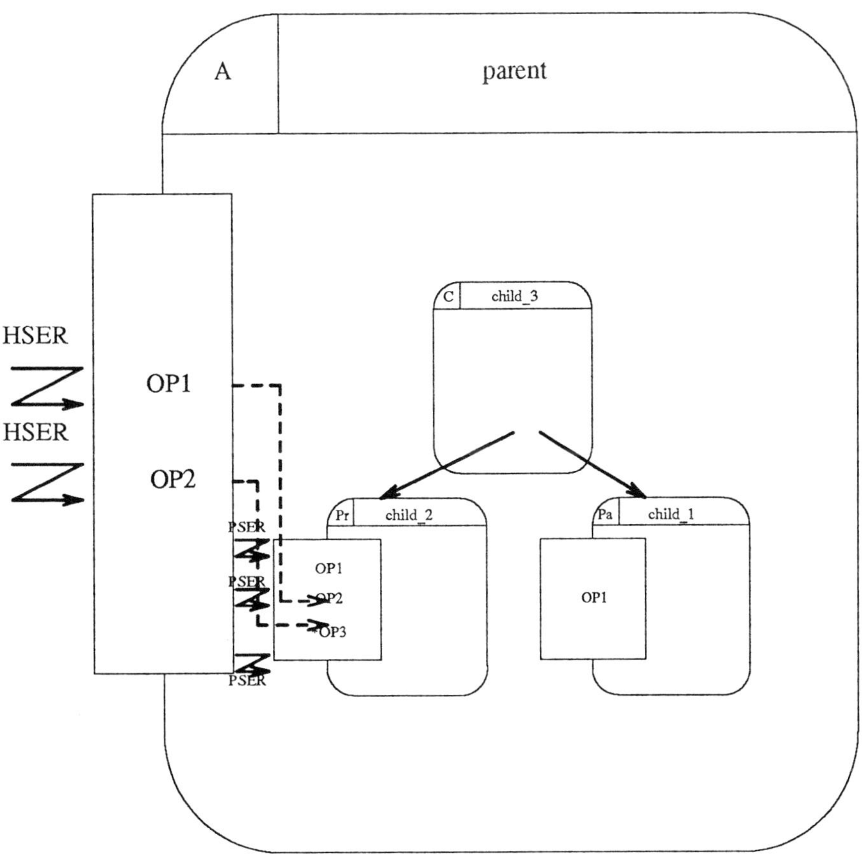

Figure 3.10: The Decomposition of an *Active* Object into *Cyclic*, *Protected* and *Passive* Objects

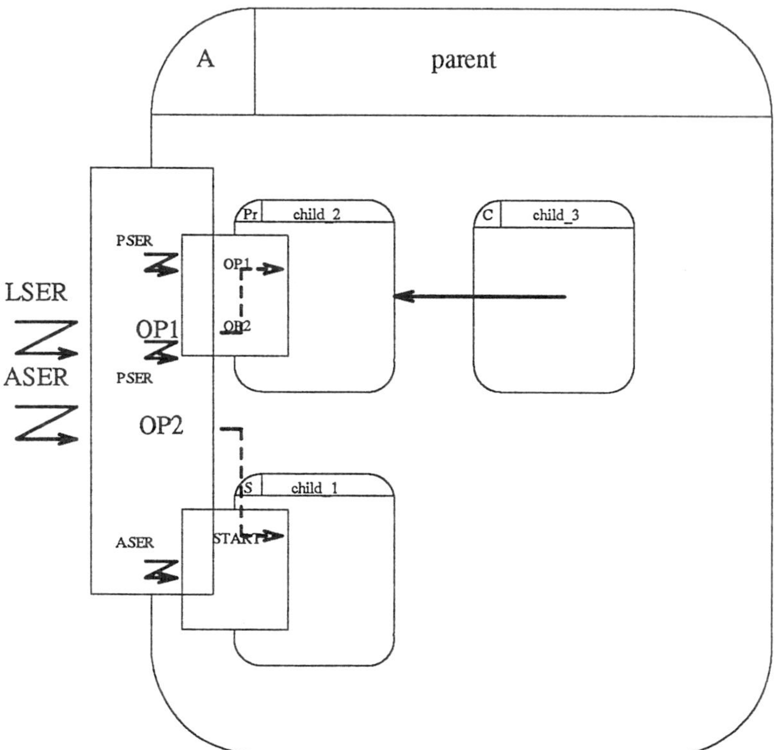

Figure 3.11: The Decomposition of an *Active* Object into *Cyclic*, *Sporadic* and *Protected* Objects

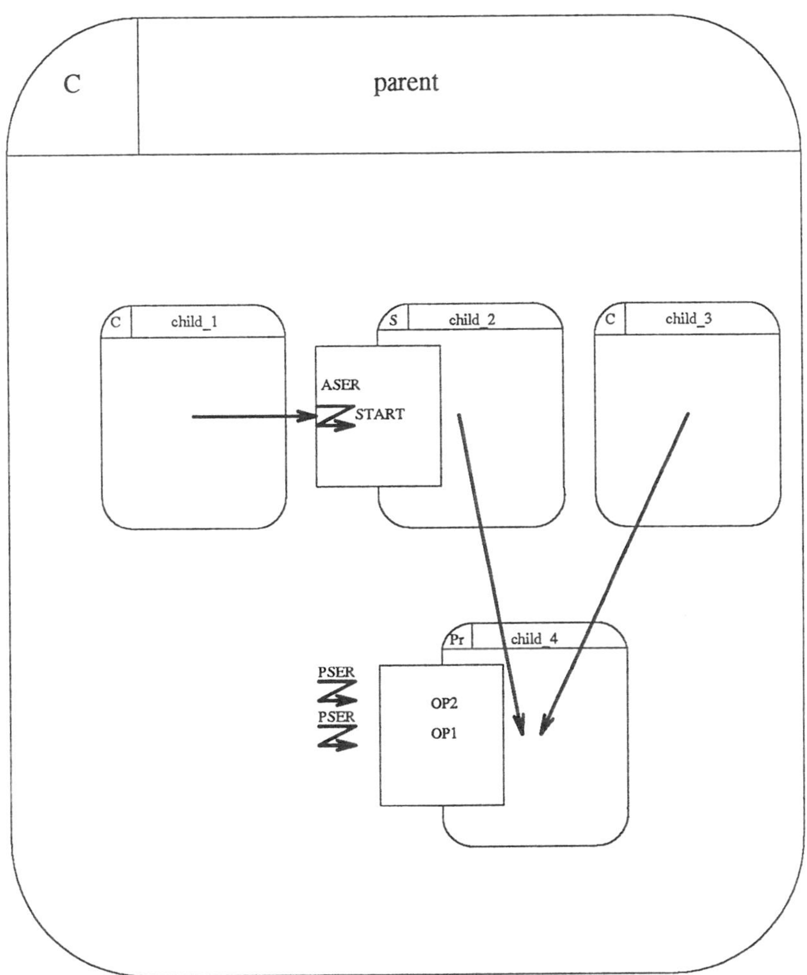

Figure 3.12: The Decomposition of a *Cyclic* Object into *Cyclic*, *Sporadic* and *Protected* Objects

3.10.3. Operation_set

An *operation_set* corresponds to a set of operations which are implemented at the next or further levels of decomposition by a set of lower level operations. Operation sets are intended as a writing facility and are represented by curly brackets ({ ... }) in the HRT-HOOD diagrams and with the keywords OPERATION_SETS in the ODS.

In the graphical representation of the parent object, the operation set will have only one "implemented by" link to one child object. Furthermore, the operations shall not decompose into other *operation sets*. The *operation set* will only be described at the object level in which each operation of the set will be identified with the keywords

MEMBER_OF <set_name> in the ODS.

3.11. Object Control Structure and Thread Decomposition

Active, protected, cyclic and *sporadic* objects all have OBject Control Structures, OBCSs. *active, cyclic* and *sporadic* objects also have *threads*.

A parent *active* OBCS must be implemented by the OBCSs of one or more child objects. The child objects must be of type: *active, protected, cyclic* or *sporadic.* The Object Control Structure field in the ODS of the parent contains the keywords IMPLEMENTED_BY and the name(s) of the child objects which implement it.

A parent *protected* OBCS must be implemented by a single child object of type *protected.* This is because the parent guarantees mutually exclusive access to its data.

A parent *cyclic* OBCS and *thread* must be implemented by a single child object of type *cyclic* or *sporadic*, or a *precedence constrained* set of *cyclic, sporadic* and *protected* objects.

A parent *sporadic* OBCS and *thread* must be implemented by a single child object of type *sporadic*, or a *precedence constrained* set of *cyclic, sporadic* and *protected* objects.

3.12. Data Flows

In order to show the flow of data, an arrow with a circle is used, with one or more data names alongside. Data may flow in the direction of use, or in the opposite direction, or both. This method is used to show the major *data flows* of a diagram (omitting error codes, etc.)

The data-flow name is an informal name which reflects in the diagram the major information exchange between objects. In the ODS a section named with the key word "DATA FLOWS" will be filled with the data-flow name expressed in the diagram.

3.13. Exception Flows

An exception is an abnormal return of control flow during execution of a provided operation. The flow of exception is thus against the normal flow of control and this is shown by a line crossing the use relationship. This line is marked with the exception name or names. In HRT-HOOD exception handling is similar to Ada where an exception is handled locally or propagated further. An exception is associated with an operation. An exception propagates along the use relationship from the operation where it is raised to the exception handler of the user object which executes the associated recovery code.

Exceptions propagated by child operations implementing parent operations, propagate also from the parent operation to its users, however, exception propagation is not shown along the implemented_relationship in the graphical representation.

In the ODS, the exceptions which propagate to a user object must be stated as part of the *provided interface* in the used object, and as part of the *required interface* in the user object.

The graphical representation of *data flows* and *exception flows* is illustrated in Figure 3.13. Note that the exception shown is raised in child_1 and propagated to child_3.

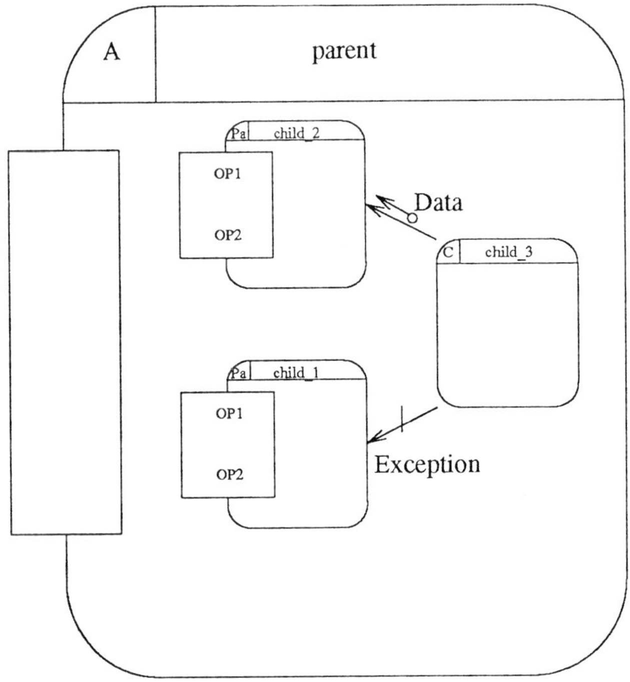

Figure 3.13: Data Flow and Exceptions in HRT-HOOD

3.14. Environment Objects

An *environment* object† represents the *provided interface* of another object (*root*) used by the system to be designed, but which is not itself part of the system. Thus an *environment* object gives the flexibility to incorporate other software into the HRT-HOOD design without impinging on the formal hierarchic decomposition principles in HRT-HOOD.

The *environment* objects may be used in the design process to allow the designer to express the context of the system to be designed (for better understanding), and to allow the interfaces between the design and its environment to be checked.

† The term "environment" originates from HOOD, and has been maintained in HRT-HOOD. Real-Time system designs often use the term environment to indicate the physical environment of an embedded computer (the process being controlled). Here the term is used in the more general sense to mean any system other than the system being designed.

An *environment* object may be required when designing an object; in this case an *environment* object will be used to provide the required external operations. It is represented by an uncle object added with the letter "E" for its object type. This *environment* object does not need to be shown at each intermediate level. This avoids adding too much complexity to the design, for those objects which are not part of the design work itself (i.e. objects which are not part of the software to be developed explicitly). A list of *environment* objects can be shown at each level of the design.

The information for the *environment* object may be supplied from another HRT-HOOD design (by exporting the interface as an *environment* object ODS), or may be entered by the designer from a document describing the interface (e.g. Interface Control Document). Alternatively, if the *environment* Object has not yet been defined elsewhere, the ODS may be created for this new object.

Clearly, the real-time properties of any environment objects must be known before schedulability analysis can be undertaken during the Physical Architecture Design activity.

3.15. Class Objects

Although HOOD (and therefore HRT-HOOD) considers itself to be object-oriented, it does not address one of the main issues that is normally associated with object oriented design/programming: namely inheritance. Instead HOOD concentrates on reuse of designs by the provision of an (generic-like) object type called a *class*†.

A *class* object may be defined to represent a reusable object in which some types and/or data, being processed by the operation of the *class*, are not fully described. An *instance* of a *class* may be made in which the type and/or data are defined explicitly, thus creating an object for a particular design.

3.15.1. Class Definition

A *class* object may be designed specifically for the system to be designed when several similar objects are needed. The class will have parameters of type and/or data, and will be *passive*, *active*, *protected*, *cyclic*, or *sporadic*.

Each *class* object is a Root object of a HOOD design. A *class* object shall not use objects of the system being designed. A list of *class* objects can be shown at each level of the system design.

The *real-time attributes*, if appropriate, of a *class* object cannot be completed until an *instance* of the *class* is created.

A *class* object may:

† We have stated early on in this report that although HOOD attempts to be language independent its roots are heavily tied to Ada. Consequently *class* objects map to Ada generics. Ada 95 will provided more support for object oriented programming, including incremental type definitions.[51] Furthermore, other target languages, such as C++, may have full inheritance models. It is beyond the scope of the current project to reflect this increased expressive power in the HRT-HOOD method.

— be decomposed into *passive, active, protected, cyclic,* or *sporadic* objects
— use only *environment* objects
— include *instances* of other *class* objects

Figure 3.14 shows a *class active* object which is decomposed into a *cyclic* and two *passive* objects. The formal parameters are shown as an uncle object attached to the object, and a link between a child and the uncle object indicates that the child required access to the formal parameters.

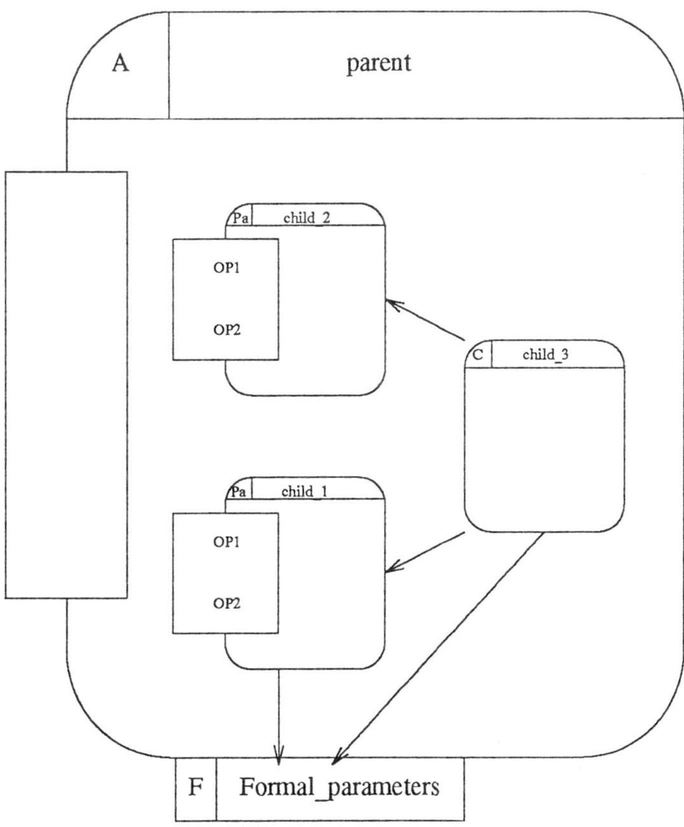

Figure 3.14: A *Class* Object in HRT-HOOD

3.15.2. Instance Definition

Instance objects will be described inside the *Design Process Tree*. Each *instance* object will be represented with a single object symbol containing the *instance* name and the *class* name. The *instance* object is declared in the INTERNALS of its parent ODS, giving the type and data parameters. The *class* object from which the *instance* object is derived, will be declared in the REQUIRED_INTERFACE of the parent ODS. An *instance* object has its own ODS.

An *instance* object of a *class* shall use only *environment* objects. When an *instance* object is created, the *environment* Objects of the *class* (if any) are added to the environment of the system being designed.

The HRT-HOOD graphical representation shows only *class instance*. For one unique *instance* object of a class the representation is an object where the name is 'typed' with the name of the class. When several identical *instances* must be represented, the names of the *instances* are generated as a generic name concatenated with the successive integer values of the index range. The graphical representation is then a double object shape with an indexed name.

When an *instance* of a *class* object is created, any *real-time attributes* of the *instance* must be completed.

3.16. Distributed Systems

The production of a distributed software system to execute on a loosely coupled collection of nodes (where a node is one or more processors sharing common memory) involves several steps which are not required when programs are produced for a single processor:

- *Partitioning is the process of dividing the system into parts suitable for placement onto the nodes of the target hardware.*
- *Configuration takes place when the partitioned parts of the program are associated with particular nodes in the target system.*
- *Allocation covers the actual process of turning the configured system into a collection of executable modules and downloading these to the sites of the target hardware.*

The HRT-HOOD method will produce a set of *terminal* objects (the unit of partitioning) which must be configured and then allocated to physical nodes in the distributed system, such that all *deadlines* are met and other constraints catered for (such as configuring and allocating device driving objects to the appropriate node). We assume here that appropriate heuristics and schedulability analysis exists to find a configuration which is feasible. For example we have used a simulated annealing approach to object configuration.[32]

The following issues need to be addressed within the HRT-HOOD framework:

1) appropriate abstractions for distributed hard real-time system design
2) configuration strategy
3) implementation strategies that can be analysed
4) mapping issues to the implementation language

In this section we will focus on issues 1) and 2). Issues 3) and 4) are considered in Part 2 of this book.

3.16.1. HRT-HOOD Abstractions

During the physical architecture design phase, *terminal objects* must be allocated to physical nodes in the distributed system. Two important issues must be considered:

a) appropriate units of partitioning, and

b) the form of communication between remote objects.

One of the main requirements of a unit of distribution is that it does not allow memory to be explicitly shared between nodes. HRT-HOOD does not constrain the type of data that can be passed between objects and therefore it is possible for one object to receive a pointer (an address) from another object. Rather than develop an additional abstraction which attempts to forbid pointers being passed between objects in a distributed environment (and to leave open the possibility of network-wide addressing), we allow them and use the presence or absence of pointers to guide the configuration process in the physical architecture design phase. Objects whose communication involves the passing of pointer will be allocated to the same node (if network-wide addressing is not supported by the execution environment).

Groups of objects may also want to be located on the same node for other reasons, for example:

- they require access to a particular physical device
- they have a tight coupling
- from a reliability viewpoint, it is easy to provide fault tolerance for all of them, as partial failures of the group can be ignored.

For the above reasons we have added an extra attribute to *non-terminal* object which indicates that all child *terminal objects* should be configured on the same physical node. The attribute is termed the *virtual node* attribute and may take an additional parameter which indicates the actual physical node for allocation. The attribute is normally set during the physical architecture design phase. The model of distribution for HRT-HOOD is therefore one in which groups of *terminal objects* are configured to nodes in the system. We do not define whether the configuration itself is represented as a HRT-HOOD design (as HOOD 3.1 does), rather we assume that some other representation may be more appropriate, depending on the configuration strategy.

Hard real-time *terminal* objects which contain *threads* (i.e. *cyclic* and *sporadic* objects) communicate with each other either through a notionally distributed *protected* object or via the "start" operation on a *sporadic*. Figure 3.15 below illustrates these approaches.

In practice, *protected* object X or *sporadic* object Y will consist of:

- the actual object X (or Y)
- a client stub for X (or Y)
- a server stub for X (or Y)

In general there are several communication mechanisms that could be supported across the network. To aid in the schedulability analysis of distributed systems we suggest that any messages sent should be asynchronous. We have, therefore, added the notion of a "protected asynchronous execution request" (PAER) to *protected* objects (these were not available in the first version of HRT-HOOD). These requests still require controlled access to the *protected* object, however, they do not necessarily require the calling object to wait for the operation to be executed (if, however, the called object is on the same machine as the calling object, the PAER is implemented as a PSER as the blocking time is bounded). It follows that with a PAER request, data can only be passed to the *protected* object (i.e. only "in" parameters are allowed). The "start" operation on a *sporadic* is defined to be asynchronous already. Requests which result in communication across the network should be allocated a *priority* during the physical architecture phase.

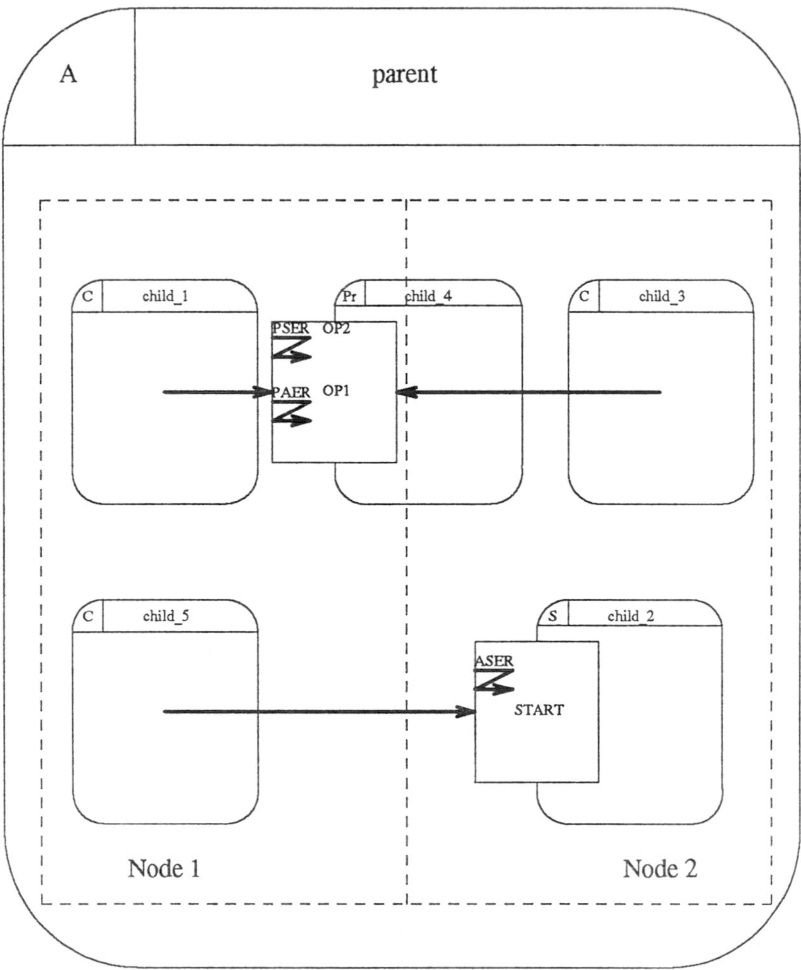

Figure 3.15: Partitioning of Terminal Objects

3.16.2. Configuration Strategy

During configuration of the *terminal* objects, objects which communicate asynchronously may be grouped on different nodes; those which communicate synchronously, or with pointers, will be grouped on the same machine.

It is not the goal of this book to define an appropriate method for performing the configuration and allocation process. However, certain constraints may be placed on that process because of the application requirements, for example:

- all *deadline*s must be met
- all resources must be utilised by not more that a set percentage
- all replicas for fault tolerance should be executed in separate nodes
- device objects must be allocated to the corresponding node of the device
- objects which require to be tightly coupled must be allocated to the same node.

3.17. Summary

This chapter has defined the basic structure of HRT_HOOD objects and has introduced their graphical representation. Each object in a HRT-HOOD design has a name, a designation, an interface that is provides for the rest of the system and a set of external (required) interfaces that the object needs. An object is designated as being either *passive*, *active*, *cyclic*, *sporadic*, *protected* or *class*. Each of these has been defined in detail within this chapter. Object attributes have also been introduced so that objects can be annotated with their real-time characteristics (e.g. *deadline*, *period* and *worst case execution time*).

There are two fundamental relationships between objects: the *include* and *use* relationships. In addition, data flows and exceptions flows can be represented. During design, an object may be subject to hierarchical decomposition so that the properties of an object at one level of abstraction are implemented by a group of objects at a lower level. This *include* relationship is concluded when a set of *terminal* objects are defined. For a hard real-time subsystem these terminal objects must be either *passive, cyclic, sporadic* or *protected.* The control flow between objects (in particular between the *terminal* objects) is represented by the *use* relationship. The HRT-HOOD approach defines the legal set of *include* decompositions and *use* relations.

Finally in this chapter the issue of distributed systems and HRT-HOOD has been considered.

Part 2
Mapping HRT-HOOD Designs to Ada

In this Part we consider the systematic translation of HRT-HOOD designs to Ada 83 and Ada 95. First, in Chapter 4 we review the support that Ada 83 and Ada 95 provide for hard real-time systems. In Chapter 5 we present a discussion of our overall approach to mapping HRT-HOOD designs to Ada. Then in Chapters 6, 7 and 8 we present the detailed mappings. Ada 95 has specified many changes to the type model of Ada 83 in an attempt to make it more object oriented. Considerations of these changes, however, are beyond the scope of this book.

It is inevitable that a restricted subset of Ada (83 or 95) will be required if a tool is to be designed that analyses Ada code for its worst case execution times. It is anticipated that this subset will *exclude* the following features:

- recursive or mutually recursive subprogram calls
- unbounded loop constructs
- dynamic storage allocation
- unconstrained arrays or types containing unconstrained arrays

Although periodic threads will generally be implemented using an Ada delay statement, the schedulability analysis cannot cope with arbitrary delays in thread execution. Consequently we *do not allow* the application programmer to use

- the delay statement.

Purpose of Mapping

The purpose of this part of the book is not to prescribe the only way in which HRT-HOOD designs can be mapped to Ada 83 or Ada 95. Rather, our goal is to illustrate how analysable code can be generated.

It should be noted that although every attempt has been made to ensure these mappings are correct, the absence of a fully validated Ada 95 compiler in particular means that errors may still exists.

4 Support for Hard Real-Time Systems in Ada 83 and Ada 95

It is widely accepted that Ada 83 lacks sufficient expressive power for programming real-time systems.[20] However, the limitations with Ada 83[73] are well understood and it is possible to engineer hard real-time systems in the language if a subset is adhered to, and if extra support is provided by the run-time environment.[46] Furthermore Ada 95 does address many of the requirements which have been formulated over the last few years by the real-time community.[67,84]

The Ada 95 language is now stable enough for practitioners to begin to consider how best to design new systems to facilitate an easy transition from Ada 83 to Ada 95. Even if new systems are not to be updated to Ada 95, given the problems with Ada 83, it is inevitable that some extra support is required from both the compiler and the run-time system. In considering the mapping of HRT-HOOD to Ada we have attempted to define some minimum changes that can be made to an Ada 83 development system (compiler and its associated run-time support system) in order to support the HRT-HOOD real-time abstractions. We have chosen these changes so that they are consistent with Ada 95 real-time abstractions. In general we are concerned with those applications which require a stand-alone Ada environment, and not those system which execute on top of a real-time operating system where alternative approaches may be more appropriate. Also, particular compiler vendors may have their own versions of these extensions which are equally suitable.

In Section 4.1 we present a brief overview of the model of real-time programming in Ada 83 and how this has been extended in Ada 95. In Section 4.2 we consider how many Ada 95 functions can be provided in Ada 83 with minimal changes to the compiler and the run-time system. Although Ada 95 has responded to most of the requirements which have been articulated by the real-time community, it does not provide any support for CPU time measuring, monitoring and enforcing. In Section 4.3 we describe how our approach can be extended to cope with this important topic. In Section 4.4 we describe our experiences with prototyping some of these changes.

4.1. The Ada 83 and Ada 95 Real-Time Models

Although Ada 83 has been found lacking in its support for real-time, there is a well-defined implementation model which underlies the facilities it does provide. This model may be summarised by:

1) Pre-emptive priority based scheduling.
2) Static priorities.
3) Synchronous communication with limited priority inheritance.
4) Synchronisation with time through access to a time-of-day (calendar) clock and a delay statement.
5) Interrupt handing via task entries.

Ada 95 has extended the Ada 83 model by supporting (either in the core language or in the Real-Time Systems Annex):

a) A larger range of priorities — in Ada 83 there was no minimum range of priorities that an implementation had to support; in Ada 95 a minimum range of 32 priority levels is required.
b) Dynamic priorities — Ada 95 allows a task to change its own priority and the priority of another task.
c) Asynchronous (data-oriented) communication — Ada 95 has introduced the notion of a protected type† which can be used to decouple interacting tasks; a protected object enables the data to be shared between tasks to be encapsulated, and operations to be defined which have automatic mutual exclusion. For single processor systems, the mutual exclusion can be implemented by allocating a "ceiling" priority to the protected object; all operations are then executed at this priority.
d) Synchronisation with a monotonically increasing clock — Ada 95 allows a task to "delay until" a time in the future, where the time can either be specified by the time-of-day clock or the monotonically increasing (real-time) clock.
e) Asynchronous transfer of control — Ada 95 has extended the select statement for entry calls (and therefore for protected entry calls) to allow a task to have its flow of control asynchronously changed by another task.
f) Interrupt handling via protected objects — Ada 95 allows a protected operation to be called directly by an interrupt.

The above mechanisms allow

† In this book we will use the term "protected object" to mean an instance of a protected type in Ada 95, the term "*protected* object" to mean the HRT-HOOD abstraction, and the term "protected task" to be the Ada 83 implementation of the protected type abstraction.

a) The assignment of priorities to tasks, for example, according to the rate monotonic or deadline monotonic scheduling theory. Previously it could not be guaranteed that an implementation would support a sufficiently large range of priorities for these priority assignments to be used.

b) The development of systems which either have modes of operation (where the priority of each task may vary between modes) or which require more dynamic scheduling, e.g., Earliest Deadline First or Least Laxity Scheduling. Previously these systems were difficult to develop due to the lack of dynamic priorities.

c) The ability to bound the blocking that a task experiences when communicating, by setting the priority of the protected objects it accesses to give immediate priority inheritance (often known as priority ceiling emulation). Ada 83 suffered from extensive priority inversion.[45]

d) A more accurate representation of a periodic task. In Ada 83, to get periodic execution the time to the next period had to be calculated and then a delay statement executed. Unfortunately this was not an atomic operation, and a task could be preempted between calculating the delay and executing the delay statement. The "delay until" statement of Ada 95 removes this race condition.

e) The ability to undertake fast mode changes etc. In Ada 83, the only way to stop a task from executing or to divert its flow of control was to abort and recreate it. In most run-times this was a costly operation. It is hoped by supporting an ATC facility directly in the language that more efficient implementations will be possible.

f) The ability to handling interrupts quickly. In Ada 83, interrupt handling was potentially very slow if the full tasking model was used; this led to the notion of fast interrupt handlers — handlers which were coded using a subset of the tasking facilities.[46] It is hoped that protected objects will provide a more efficient interrupt handling mechanism.

4.2. Supporting Ada 95 Abstractions in Ada 83

Most of the extra real-time features for Ada 95 have been added in order to increase the functionality and efficiency of the run-time support system. Traditionally in Ada 83 corresponding facilities have been provided by packages whose bodies are linked with the run-time system.[46] In this section we consider how many of the Ada 95 facilities can be implemented in Ada 83 with minimum changes to the compiler and run-time.

4.2.1. Support for Preemptive Priority-based Scheduling

For most run-time support systems, increasing the range of priorities to be supported is a relatively simple matter. Furthermore ordering entry queues and a priority driven select statement can easily be added. However, if the application area is using only protected tasks for communication and synchronisation (see Section 4.2.3) and not the generalised rendezvous primitives, then a priority select is not required, priority entry queues are only required if more than one task can be queued on a protected task entry.

4.2.2. Support for Dynamic Priorities

In Ada 95, a task can query and alter the priority of another task. This requires a mechanism of accessing a unique identifier of a task. In Ada 95, this is achieved using a package defined in the Real-Time Systems Annex. In Ada 83, the run-time support systems can provide a similar mechanism. We adopt the following simplified version of the CIFO[46] entry for task identifiers — the difference being that for hard real-time systems we are not concerned with task hierarchies and therefore do not provide mechanisms by which a task can find its parent, master or children.

```
package TASK_IDS is

  type TASK_ID is private;

  TASK_ID_ERROR : exception;

  function NULL_TASK return TASK_ID;

  function SELF return TASK_ID;

  function CALLER return TASK_ID;

  function CALLABLE(I : TASK_ID ) return BOOLEAN;

  function TERMINATED(I : TASK_ID ) return BOOLEAN;
private

  type TASK_ID is <IMPLEMENTATION-DEFINED>;

end TASK_IDS;
```

This package provides the type TASK_ID and several functions for obtaining the identifier of a particular task and determining its status. Most of the routines are self explanatory. The function CALLER, if called from within an accept, returns the ID of the partner task, which made the entry call which is being accepted. If called from outside an accept, it returns NULL_TASK. The exception

TASK_ID_ERROR is defined but not raised directly by this package. The support for dynamic priorities raises this exception.

```
with SYSTEM; use SYSTEM;
with TASK_IDS; use TASK_IDS;

package DYNAMIC_PRIORITIES is

  procedure SET_PRIORITY(T: TASK_ID; TO : PRIORITY);

  function GET_PRIORITY(T: TASK_ID) return PRIORITY;

end DYNAMIC_PRIORITIES;
```

The DYNAMIC_PRIORITIES package provides support for querying and altering the base priority of a task. The effect of the SET_PRIORITY procedure is to cause the base priority of the specified task to be set to the specified value. If the task is terminated, the operation has no effect. The GET_PRIORITY function returns the current base priority of the specified task. If the task is terminated, the operation has no effect. The exception TASK_ID_ERROR is raised if the task identifier is null.

4.2.3. Support for Immediate Ceiling Priority Inheritance

In Ada 95, tasks can communicate and synchronise via protected objects. These are entities which encapsulate state and have a well defined interface. Access to a protected object is guaranteed to be in mutual exclusion (if the Real-Time Systems Annex is supported) and to use immediate priority ceiling inheritance to bound the blocking time. Protected objects do not have an internal thread of control. Ada 95 protected objects can be used to implement the HRT-HOOD notion of a *protected* object. To implement a *protected* object in Ada 83 it is necessary to use a task (called a protected task). All other tasks in the system (those found in *cyclic* and *sporadic*) objects communicate with each other via protected tasks. The structure of a protected task is given below:

```
task PROTECTED_TASK is
  pragma PRIORITY(CEILING);
  pragma PROTECTED_TASK; -- recognised by compiler
  entry OP1(...);
  entry OP2(...);
end PROTECTED_TASK;
```

```
task body PROTECTED_TASK is
  -- no local variables
begin
  loop
    select
      when G1 =>
        accept OP1(...) do
          ...;
        end OP1;
    or
      accept OP2(...) do
        ...;
      end OP2;
    or
      terminate;
    end select;
  end loop;
end PROTECTED_TASK;
```

Protected tasks cannot have delay statements in them (or other potentially blocking operations); they have ceiling priorities assigned†.

The implementation of protected tasks can be undertaken in two ways:

(1) As an ordinary (real) server task.

(2) As a passive task‡, if pragma PASSIVE is used and such an optimisation is supported by the compiler and run-time system; that is, there should be no run-time thread associated with the task; the operations should be executed on the stack of the calling tasks using the priority mechanism (on a single processor system) to provide mutual exclusion. If possible the compiler should provide the same semantics as if the task were a protected type in Ada 95.

If an ordinary task is used simply to give mutual exclusion over some resource (there are no guards) then, as a result of its ceiling priority and lack of delay statement, it cannot have more than one client task on its queue. Hence the issue of FIFO queues is not a problem (i.e. Ada 83 standard implementation are usable). After a client task has completed a rendezvous, the server task (which must have a higher priority) will loop round and re-execute the select statement (or simple accept). The time for this code is added to the cost of the client's call.

†The scheduling analysis tests do not consider these tasks directly. Only the client tasks are analysed, although the time spent in each protected task must be known for the calculation of WCET and blocking.

‡ Many Ada 83 compilers already support the notion of a "passive" task. Passive tasks usually control access to shared data or are used to provide fast interrupt handlers, and therefore do not require an independent thread of control. Passive tasks are typically indicated by a pragma and the compiler will generate specific calls to the run-time systems.

Note that if guards are used then current scheduling theory can not deal with arbitrary condition synchronisation. If, for example, a finite buffer is used to link two tasks then the need to block the producer, if the buffer is full, means the program cannot be analysed without semantic information about the application. In hard real-time systems it may be better to make the producer non-blocking (i.e. overwrite a full buffer) and construct the consumer as a sporadic that is activated every time there is some object in the buffer. Arbitrary synchronisation is allowed for the implementation of mode changes; here each mode will have to be analysed separately and the change over period will also have to be analysed.[81]

With the current definition of the semantics of protected types in Ada 95, a client task when it leaves the protected object will execute the protected code on behalf of any blocked client that can now proceed (because the original task has changed the state of the object). The time for this extra *epilogue* code must be added to the WCET for each client task. If passive tasks are supported, then Ada 95 protected type semantics can be implemented†.

4.2.4. Support for Periodic Task Execution

We have already mentioned the requirement for a "delay until" construct. It is possible to provide the following simple run-time package which provides the same semantics.

```
with CALENDAR; use CALENDAR;
package CALENDAR_DELAY_SUPPORT is

  procedure DELAY_UNTIL(T : TIME);

end CALENDAR_DELAY_SUPPORT;
```

Therefore a periodic task can take the form:

---

†Extra pragmas will need to be added to the compiler to help any WCET analyser tool calculate the maximum time spend in the epilogue code. For example the following pragmas could be added:

- Pragma MAX_QUEUED(entry_name, integer);

  This pragma can be applied to any guarded entry and indicates the maximum number of tasks which could be queued on the entry at anyone time. This information can be supplied by the user or can be found from the HRT-HOOD design. Typically for sporadic tasks the value is 1.

- Pragma RELEASE(entry_name, protocol);

  This pragma can be applied to any guarded entry and indicates that when the guard is set to true, then either ALL the tasks should be release or just ONE. Note that if just ONE task is released then the entry queues should be priority ordered.

Although the above approach does not allow the programmer to express algorithms which release a variable number of tasks (for example some resource control algorithms), we contend that it will be adequate for most hard real-time systems.

```
declare
  NEXT : TIME;
  INTERVAL : constant DURATION := ...;
begin
  NEXT := CLOCK + INTERVAL;
  loop
    -- code to be executed
    DELAY_UNTIL(NEXT);
    NEXT := NEXT + INTERVAL;
  end loop;
end;
```

In practice, however, many embedded systems are not concerned with calendar time but use the notion of monotonic time (usually the number of time units since system initialisation). Hence a "delay until" function is required which takes monotonic time. The following packages will suffice.

```
package REAL_TIME is

  type TIME is private;

  function CLOCK return TIME;

  function "+" (LEFT: TIME; RIGHT: DURATION) return TIME;
  function "+" (LEFT: DURATION; RIGHT: TIME) return TIME;

  function "-" (LEFT: TIME; RIGHT: DURATION) return TIME;
  function "-" (LEFT: TIME; RIGHT: TIME) return DURATION;

private
  type TIME is -- implementation dependent;
end REAL_TIME;

with REAL_TIME; use REAL_TIME;
package REAL_TIME_DELAY_SUPPORT is

  procedure DELAY_UNTIL(T : TIME);

end REAL_TIME_DELAY_SUPPORT;
```

Note, however, that where the granularity of duration is too coarse it may be appropriate to introduce a new type (called `Time_Span` in Ada 95).

4.2.5. Support for Asynchronous Transfer of Control and Mode Changes

Asynchronous transfers of control are generally required for emergency situations, such as unanticipated mode changes, error conditions etc. Although Ada 83 does not possess an asynchronous transfer of control facility, it can be argued that such a facility can be efficiently provided using the Ada abort.[21,46] The model adopted here is that a task which wishes to be interrupted declares an entry which will be called when the interruption is to occur. An agent task executes the code which is to be interrupted. The following example illustrates the approach.

```
task SERVER is
  entry START_COMPUTATION(<PARAMETER PART>);
  entry ATTENTION(<PARAMETER PART>);
  -- above entry called by client

  entry AGENT_STARTED(<PARAMETER PART>);
  entry AGENT_FINISHED;
  -- these entries called by agent

end SERVER;

task body SERVER is

  task type AGENT;
  type ACCESS_AGENT is access AGENT;

  AGT : ACCESS_AGENT;

  task body AGENT is
  begin
    loop
      SERVER.AGENT_STARTED(<PARAMETER PART>);
      -- computation
      SERVER.AGENT_FINISHED;
    end loop;
  end AGENT;
```

```
begin
  loop
    AGT := new AGENT;

    accept START_COMPUTATION(<PARAMETER PART>) do
        accept AGENT_STARTED(<PARAMETER PART>) do
            -- pass parameters to agent;
        end AGENT_STARTED;
    end START_COMPUTATION;

    -- select agent completion or call to ATTENTION
    select
      accept AGENT_FINISHED;
      accept ATTENTION(<PARAMETER PART>) do
        -- return full results
      end ATTENTION;
    or
      accept ATTENTION(<PARAMETER PART>) do
        abort AGT.all;
        -- return any partial results
      end ATTENTION;
    end select;

  end loop;
end SERVER;
```

Unfortunately, this approach will require that variables be shared between the created agents. This will require the use of PRAGMA shared and dynamic memory management. Ideally this should be optimised so that no dynamic memory management is required. We will use a stylised version of this approach in the mapping of *sporadic* and *cyclic* objects.

Although in general a compiler and run-time system could be modified to recognise the above structures, in practice this is likely to be non-trivial. Consequently it is not easy to bridge the gap between Ada 83 and Ada 95 in this area. Furthermore, it is still an open issue as to whether such an approach can produce code which can be analysed for its worst case execution time.

4.2.6. Support for Interrupt Handling

In Section 4.2 we indicated that the Ada 95 approach to interrupt handling is to allow interrupts to call a procedure in a protected object. Given the implementation of protected tasks described above it is relatively simple to allow an address clause to be placed on an entry.

4.3. Extending the Model

Although Ada 95 has addressed directly most of the real-time requirements there is one major topic which is currently deemed out-of-scope; this is associated with measuring, monitoring and enforcing the execution time of a task.[29] In this section we consider how this facility can be provided by interfacing into the run-time system. Given that this facility is not provided by Ada 95 we present a brief discussion on its utility.

4.3.1. Measuring and Monitoring and Enforcing CPU Execution Time Usage

Many real-time scheduling theories depend on the ability to measure or estimate the amount of CPU time tasks require. For example, the schedulability tests for rate monotonic scheduling[64] require that the worst case execution time (WCET) of each task be known. There are three approaches by which the execution time requirements of a task can be obtained. The first is testing, which is a method of evaluating the timing characteristics of code fragments by actually running them on representative hardware with appropriate test data. A disadvantage of this is that test data may not fully cover the domain of interest. Its advantage is that it allows the programmer to determine the typical cost of executing perhaps complex and unbounded algorithms. Second, the timing properties of the program can be analysed by simulating the target system. Such methods rely heavily on the model of the underlying hardware, which is an approximation of the actual system, it may therefore not accurately represent the worst case situation. The third, is to estimate the worst case program execution time by analysing the programs at either high language level, or both high and assembler language levels.

Having estimated or measured the CPU time required by each task, many real-time techniques require that the execution time actually consumed at run-time be monitored. In general there are three reasons why this might be the case.

- Timing Fault Tolerance

  A system can no longer be guaranteed to meet all its deadlines if one or more of its tasks over-run their CPU allocation. In this case there are several options, including the following.

  — Allow the task to over-run and hope that other tasks still make their deadlines. The advantage of this approach is that if not all tasks need their worst case execution time, no deadlines may be missed. If deadlines are missed then the system can be designed so that non-crucial tasks miss their deadlines first. The disadvantage is that the task that suffers the missed deadline is not the task that caused it.

  — Do not allow the task to over-run. This approach requires that a task be notified if it over-runs its execution time. This has the advantage that error handling can be performed by the task in error, and no other deadlines are missed.

— A combination of the above, e.g. allow non-crucial tasks to over-run but not crucial ones, and ensure that no over-run can result in a crucial task missing its deadline.

- Period Transformations

  With priority-based scheduling schemes it is often required to transform the execution of a periodic task into small units so that other tasks may be scheduled to meet their deadlines, i.e. lower priority tasks do not suffer such a large preemption (in the worst case).[70] One way of achieving this is to allow a task to execute until it has consumed a certain amount of CPU time and then lower its priority (or suspend it) thus allowing other tasks to execute. Later its priority may be raised again (or it may be resumed).

- Sporadic Servers

  In some systems it is difficult to bound the minimum inter-arrival time of aperiodic events, and consequently a mechanism must be found by which they can be handled within hard real-time systems. This can be done in a number of ways. The simplest approach is to provide a periodic task whose function is to service one or more aperiodic tasks. This periodic-server task can be allocated the maximum execution time commensurate with continuing to meet the deadlines of the periodic tasks. As aperiodic events can only be handled when the periodic server is scheduled, the approach is, essentially, polling. The difficulty with polling is that it is incompatible with random arrival times. When the server is ready there may be no task to handle. Alternatively the server's capacity may be unable to deal with a concentrated set of arrivals. To overcome this difficulty a number of *bandwidth preserving* algorithms have been proposed: Priority Exchange,[62,77] Deferrable Server[62,71,78] and Sporadic Server.[62,71,78] All of these algorithms involve allocating a fixed budget of CPU time to each aperiodic event. When these budgets are exhausted the task must be suspended until more budget has been generated.

It is clear from the above discussion that CPU execution time usage is an important parameter in the design and construction of hard real-time systems.

Ada 83 has no concept of measuring execution time, an early Ada 95 mapping did address the issue but its is currently viewed as being out-of-scope. Ada 95 does allow a missed deadline to be detected using the ATC facility (the "select and then abort" facility with a delay branch will provide deadline overrun detection).

4.3.2. Support for CPU Execution Time Monitoring

The following package is an example interface into the run-time system to support CPU execution time monitoring and enforcing. It assumes hardware support for a count-down clock with a granularity finer than the normal time-of-day clock. If the real-time clock is of sufficent granularity, then the type `Time_Span` can be used. The clock generates an interrupt when the value of a counter reaches zero.

```
package CPU_BUDGETS is

   type BUDGET_ID is private;
   -- assume the following is either declared here or in package
   -- REAL_TIME
   type TIME_SPAN is <IMPLEMENTATION DEPENDENT>;
     -- say delta 1.0 range 0.0 .. 60.0;
   for TIME_SPAN'SMALL
       use <IMPLEMENTATION DEPENDENT>;
     -- for example, 0.000004 (250 KHZ interval timer)
   NULL_ID : constant BUDGET_ID;
   BUDGET_OVERRUN: exception;
   -- raised if budget time exceeded
   BUDGET_SETTING_ERROR: exception;
   -- raised if a non active budget id is passed to CANCEL

   function SET_BUDGET(LIMIT : TIME_SPAN)
                                 return BUDGET_ID;
   function TIME_LEFT return TIME_SPAN;
   function LAST_OVERRUN return BUDGET_ID;
   -- of the calling task

   procedure CANCEL(B : in out BUDGET_ID);
   -- B is set to NULL_ID
   -- CANCEL(NULL_ID) is a Null operation
private
   type BUDGET_ID is new INTEGER;
   NULL_ID : constant BUDGET_ID := 0;
end CPU_BUDGETS;
```

The paradigm for measuring a section of code is:

```
with CPU_BUDGETS; use CPU_BUDGETS;
   ...

MY_ID : BUDGET_ID;
TIME_ALLOWED : TIME_SPAN;
   ....
```

```
begin
  MY_ID := SET_BUDGET(TIME_ALLOWED);
  -- section of code to be timed
  CANCEL(MY_ID);
exception
  when BUDGET_OVERRUN =>
    if MY_ID = LAST_OVERRUN then
      -- recovery algorithm
    else
      CANCEL(MY_ID);
      -- we are in real trouble
      raise;
    end if;
  when BUDGET_SETTING_ERROR =>
    -- not properly nested budget request
    ...;
end;
```

Note that this, in effect, raises an asynchronous exception in this task. The semantics of this should be as if an agent task was created and then aborted.

There are several problems which arise when attempting to implement this approach. They all stem from the having to deal with nested budgets:

1) an inner budget may be set with a larger value then the time outstanding on an outer budget;

2) if an outer budget expires, then the inner budget may expire whilst in the budget exception handler (and therefore will be lost);

3) two budgets may expire simultaneously (hence giving the problem of multiple asynchronous exceptions).

To solve these problems we note that as budgets represent worst case execution times, an inner budget should not be greater than an outer budget. We can therefore define the following restriction: if an inner budget is set which is greater than any currently outstanding budget for that task (by some value D which is approximately equal to the time spent in the exception handler of the inner budget), then a run-time error is raised at budget setting time. This ensures the required behaviour of the budget mechanism.

With an Ada 95 implementation, the `CPU_BUDGETS` package would probably be represented as a child package of `Ada.Real-Time`.

4.4. Implementation Cost

All of the functions except dynamic priorities and asynchronous transfer of controls have been prototyped by modifying an existing Ada run-time support system based on the YSE Ada Compiler.[1] Although we do not anticipate difficulties in implementing dynamic priorities, ATC would require much more effort. In general the modifications that have been undertaken have been relatively simple and have taken of the order of a few person-months of effort. Not surprisingly it took more time to test the new kernel, undertake a performance analysis,[30] and produce appropriate documentation than it took to implement the changes.

Our initial experiences with using the kernel have been positive. However there are several observations that should be made.

Protected Tasks

Although the implementation of a protected task produced an improvement in performance compared to a standard rendezvous it was not optimised for interrupt handling and we were not able to get sufficient performance to use the mechanism for handling fast interrupts off a communications bus.[27,31] It was clear from our experiments that for really fast interrupts it was necessary to associate an Ada procedure with the interrupt. Consequently this facility was added to the compiler and run-time system.

Delay Until

In Section 4.2 we indicated that a "delay until" abstraction when used in conjunction with package calendar gave a more accurate representation of a periodic task. In practice the overhead involved with working with package calendar is too great for many application domains: to implement a periodic task involves the following:

```
NEXT := CLOCK + INTERVAL;
loop
  -- code to be executed
  DELAY_UNTIL(NEXT);
  NEXT := NEXT + INTERVAL;
end loop;
```

Although this appears quite simple, the implementation of "delay until" is

```
-- call the run-time to turn clock interrupts off

delay(CLOCK - T); -- where T is the parameter passed

-- call the run-time to place the task in the delay
-- queue and turn interrupts back on
```

Unfortunately subtracting two calendar time values is a non-trivial operation as (in the general case) these values may involve extended periods of time with leap years in between. In our implementation this subtraction took a few milliseconds. Consequently package "real_time" should be used and it can give a more appropriate response time for time critical tasks.

4.5. Summary

The definition of Ada 95 is stable. Unfortunately it could be 2-3 years before fully validated compilers become available which support the real-time facilities. In general Ada 95 has been designed to be upward compatible with Ada 83, furthermore many of the new real-time facilities can be provided by an extended Ada 83 compiler and run-time system. In developing the mappings to HRT-HOOD, we have defined a subset of these extensions which we believe offers the user community significant improvement on Ada 83 and yet will provide transition to Ada 95 with minimal syntactic changes.

5 Overall Mapping Approach

The code generation rules for HOOD 3.0 to Ada 83 [3] were, informally, widely criticised for being too inefficient and for generating too many Ada compilation units. The structure of the previous mappings for HRT-HOOD were based on the HOOD 3.0 mappings[24] and therefore open to the same criticism. Although HOOD 3.1 has moved away from giving explicit mappings to Ada, it does present rules for translating into Ada.[4] This section considers these rules and gives a simple example of the type of code that might be generated if the HOOD 3.1 approach was adopted. It illustrates that code generation is still not as efficient as it could be, and proposes an alternative mapping for HRT-HOOD. We shall assume an Ada 83 package structure for these discussions.

5.1. HOOD 3.1 to Ada 83 Mapping

In this section we consider the mapping of a simple *passive* object to an Ada 83 package structure. This enables us to present an overview of the proposed scheme without getting involved in details of concurrency.

In HOOD 3.1, a *passive* objects maps to a single Ada package (in HOOD 3.0, there were potentially several Ada packages: one for the provided types, one for the internal types and data, and one for the object itself). Operations, by default, are represented as subunits. For terminal objects the mapping is simple. For example, a passive object which declares a type, a constant, an exception and two operations would generate the following Ada code:

```
package OBJECT_NAME is
  type X is new INTEGER range 1..10;

  C : constant X := 3;

  E : exception;

  procedure OP1;

  function OP2 return X;
end OBJECT_NAME;

package body OBJECT_NAME is
  -- internal types, constants,  operations etc

  procedure OP1 is separate;

  function OP2 return X is separate;
end OBJECT_NAME;
```

This mapping is simple and efficient.

Consider now a parent *passive* object which is decomposed into two child objects, where each object implements one of the parents provided operations etc. The parent object mapping suggest by the HOOD 3.1 extraction rules is given by:

```
with CHILD1, CHILD2;
package PARENT_SERVER is
  subtype X is CHILD1.X;

  C : X renames CHILD1.C;

  E : exception renames CHILD2.E;

  procedure OP1 renames CHILD2.OP1;

  function OP2 return X renames CHILD1.OP1;
end PARENT_SERVER;
```

```
package CHILD1 is
  type X is new INTEGER range 1..10;

  C : constant X := 3;

  function OP1 return X;
end CHILD1;

package CHILD2 is
  E : exception;

  procedure OP1;
end CHILD2;
```

The main concern with this mapping is that it generates redundant packages; every non-terminal object in the system has a package specification whose contents simply rename types, object etc to the child objects which implement them. Furthermore, all the non-terminal package specifications have to be elaborated at run-time.

The motivation for this mapping is the terminal object visibility rules of HOOD; a particular terminal object may not have direct visibility to the terminal object which provides the operations it requires, but only to the object's parent. So, for example, a client of the above packages might have the following code generated for it.

```
with PARENT_SERVER;
package CLIENT is
  procedure OP1;
end CLIENT;

package body CLIENT is
  procedure OP1 is separate;
end;

separate(CLIENT)
procedure OP1 is
begin
   ...
   PARENT_SERVER.OP1;
end;
```

There is a further point to be noted with this mapping in that the code for the parent cannot be generated until the code for all its children have been generated. Therefore it is not possible to develop the client object and simply generate code for the client until the parent server has been completely decomposed (although one could imagine generating a temporary package for the parent which contained all the type information etc). This problem was also manifest in HOOD 3.0 and the original version of HRT-HOOD.

5.2. An Alternative Translation Approach

Each non-terminal object in HOOD (and HRT-HOOD) has an interface defined in the ODS. It is from this interface that the package specification for the object can be generated. Given that in order to generate this specification it is necessary to have at least the specification of all child objects, it is possible for the code generation tool to resolve all the calls directly to the child object. So, the example given above could potentially generate the following code for the client:

```
with CHILD1, CHILD2; -- children of parent_server;
package CLIENT is

  procedure OP1;
end CLIENT;

package body CLIENT is

  procedure OP1 is separate;

begin
  null;
end;

separate(CLIENT);
procedure OP1 is
begin
   ...
   CHILD2.OP1; -- for parent_server.op1;
end;
```

The arguments in favour of this approach are:

1) it cuts down on the number of redundant Ada compilation units by eliminating all code generated by non-terminal objects

2) it produces more efficient run-time code by cutting down the number of library units to be elaborated

3) it allows for potential optimisations; for example, the client above only needs to "with" CHILD1; the code for OP1 can be generated in-line.

The disadvantages are:

a) it requires the code generator to recognise a call in the code to "parent.op1" and to resolve this call to "child1.op1"; each call to be resolved would perhaps have to be tagged by the designer, for example "$parent.op1$" (the same would apply to all types, exceptions etc)

b) there might be some loss of traceability between the generated code and the code given in the ODS; although comments could be inserted by the code generator to indicate the resolution.

When generating code for the client, if the server has not been fully decomposed into its children, the tool could generate temporary parent specification which would be deleted when the children had been fully developed.

5.3. Mapping HRT-HOOD to Ada

In this book, we shall adopt the approach whereby no code is generated for non-terminal objects. The translation to Ada consists of two stages:

— the generation from the Object Description Skeletons of the basic program structure incorporating the OPCS, thread and OBCS code; this generation can easily be automated

— the refinement of operations and object control structures from the pseudo code into Ada 95 by the designer/programmer.

In general, an object is mapped to a package, and an operation is mapped to a subprogram. The OBCSs are mapped either to an Ada protected task (in Ada 83) or to a protected type (in Ada 95), and each thread is mapped to a task. Table 5.1 shows the detailed mapping of each ODS entry in a terminal object to Ada.

Table 5.1: ODS to Ada	
ODS entity	Ada entity
Object definition	
Object_name	package name
PASSIVE	package
ACTIVE	package with a protected task/type and one or more tasks
PROTECTIVE	package with a protected task/type
CYCLIC	package with a protected task/type and a task
SPORADIC	package with a protected task/type and a task
ENVIRONMENT	package specification only
CLASS	generic package
INSTANCE	instantiation of a generic package
FORMAL_PARAMETERS	parameters of a generic package
Description	
DESCRIPTION	comments in the package specification
Real-time attributes	Real-time attributes file plus:
CEILING_PRIORITY	Ada pragma for a protected task/type priority
PERIOD	access to a real-time clock and a delay until statement
MIN_ARRIVAL_TIME	comment in package specification
MAX_ARRIVAL_ FREQUENCY	comment in package specification
DEADLINE	comments in the package specification
PRIORITY	priority pragma in a task specification
IMPORTANCE	comment in the package specification
WCET	comments in the package specification
BUDGET	comments in the package specification, use of CPU_BUDGET package

Table 5.1: ODS to Ada	
ODS entity	Ada entity
Implementation constraints	
IMPLEMENTATION_OR_ SYNCHRONISATION_ CONSTRAINTS	comments in the corresponding object's package specification
Provided Interface	
TYPES†	types in package specification
CONSTANTS	constants in package specification
OPERATIONS	subprograms in package specification
OPERATION_SET	the operations which are members of the set are declared as subprograms in the package set
EXCEPTIONS	exceptions in package specification, and comments giving operation to exception mapping
Required Interface	
FORMAL_PARAMETERS	generic formal parameters
OBJECTS	WITH clause in the package specification
TYPES	-- comment in package specification
CONSTANTS	-- comment in package specification
OPERATION_SETS	-- comment in package specification
OPERATIONS	-- comment in package specification
EXCEPTIONS	-- comment in package specification
Dataflow and Exceptionflow	
DATAFLOW	none
EXCEPTIONFLOW	none

† For terminal objects if a type is declared as PRIVATE, its declaration will be declared as private.

Table 5.1: ODS to Ada	
ODS entity	Ada entity
Object control structure	
DESCRIPTION	comments in a protected task/type specification
CONSTRAINED_OPERATION	protected task/type entry, or protected type procedure
CONSTRAINED_BY	label may become representation clause
CODE	OBCS protected task/type body
Internals	
OBJECTS	not present in terminal object
TYPES	type declarations in package body
DATA	data declarations in package body
CONSTANTS	constant declarations in package body
OPERATION_SETS	subprograms within the package set body
OPERATIONS	separate subprogram declaration in package body
EXCEPTIONS	exception declarations in package body
Operation control structure	
OPERATION	separate subprogram declaration
DESCRIPTION	comment in separate subprogram
USED_OPERATIONS	use clause in subprogram
PROPAGATED_EXCEPTIONS	comment in subprogram
HANDLED_EXCEPTIONS	Ada exception handler in subprogram
CODE	code content of separate subprogram

The packages which can be generated by an object <Name> during its translation into Ada are:

<NAME> -- the code for the object
<NAME_RTATT> -- definition of the real-time attributes for the object

In the following sections:

- <parameter_part> is defined in Appendix B — the BNF form of the ODS
- <OPCS_CODE> and <OBCS_CODE> represent the contents of the CODE sections of the OPCS and OBCS respectively

- HSER, LSER, ASER, PAER, PSER, ASATC, LSATC, HSATC are "type of request" constraints as defined in Chapter 3.

The translation process does often require several "with" and "use" clauses to be generated in order to gain access to the required interface. For conciseness these will be shown as

```
with REQUIRED_INTERFACE;
```

Real-Time Attributes

The structure of the package giving an object's real-time attributes is given below. Although the data in this package could be incorporated into the package representing the object, we prefer to keep it separate so that analysis tools can potentially access it.

First it is necessary to define some system-wide types and objects. These should be seen as part of the HRT-HOOD infrastructure and not part of the system being designed.

```
with SYSTEM; use SYSTEM;
with REAL_TIME; use REAL_TIME; -- or 95 equivalent
with CPU_BUDGETS; use CPU_BUDGETS; -- or 95 equivalent
package RTA is

  -- the following types are assumed by the mapping

  type SYSTEM_WIDE_MODE is range 1 .. 10;
    -- This type indicates the number of system-wide
    -- modes; each object may have its own internal modes
    -- of operation which is concern to the application'S
    -- programmer. If a mode of operation has a global
    -- effect on the system, for example it results in a
    -- priority change, then it should be declared global

  type IMPORTANCE is (HARD, SOFT, BACKGROUND);
    -- or some other appropriate application terms, such as
    -- safety_critical, mission_critical, or non_critical

  type OPERATION_ATTRIBUTES is
    record
      BUDGET  : TIME_SPAN;  -- allocated time
      WCET    : TIME_SPAN;  -- + error handling
    end record;
```

```
    type THREAD_ATTRIBUTES is
      record
        PRI      : PRIORITY;
        TRANS     : TIME_SPAN;   -- transformation factor
        IMP      : IMPORTANCE;
        DEADLINE : TIME_SPAN;
        BUDGET    : TIME_SPAN;   -- allocated time
        WCET      : TIME_SPAN;   -- + error handling
      end record;

    -- maximum arrival frequency  for sporadic
    type FREQUENCY is
      record
         NUMBER : POSITIVE;        -- invocations in a
         IN_PERIOD : TIME_SPAN; -- certain period
      end record;

     -- the following are assumed by the mapping

     DEFAULT_PRIORITY : constant PRIORITY := 1;
        -- set to some low value


     START_MODE : constant SYSTEM_WIDE_MODE := 1;
        -- which mode the systems should starts in

     CURRENT_MODE : SYSTEM_WIDE_MODE;
        -- current system wide-mode of operation

    SYSTEM_START_UP_TIME : TIME := CLOCK;
        -- the time at which the system started, used
        -- to coordinate tasks with offsets
  end RTA;
```

There are two points to note about this package:

1) In general as an object can exists in several modes of operation, the real-time attributes of an object is kept on a per mode basis. Some of this information(for example: priority etc) may not be usable with Ada 83 unless some of the extensions discussed in Section 2 are supported. If budget time is not supported then it can be omitted. If dynamic priorities are not supported then a single priority that is the maximum over all modes must be used.

2) In a distributed system this package must be replicated on each node. We do not define whether the clock are synchronised. However, if "offsets" are being used to order tasks across the network then some form of

synchronised global time base will need to be supported (see Chapter 8 for a discussion on distributed system issues).

The following package illustrates how an object's real-time attributes are represented in Ada.

```
with RTA; use RTA;
with SYSTEM; use SYSTEM;
with REAL_TIME; use REAL_TIME;
package NAME_RTATT is
   -- for each operation
   OP_NAME : array(SYSTEM_WIDE_MODE) of constant
             OPERATION_ATTRIBUTES := ( .. );

   -- for protected, sporadic, cyclic and active objects
   CEILING :  array(SYSTEM_WIDE_MODE) of  constant
             PRIORITY := ( .. );
   INITIAL_PROTECTED_PRIORITY : constant PRIORITY := ...;

   -- for each cyclic and sporadic object
   THREAD : array(SYSTEM_WIDE_MODE) of constant
             THREAD_ATTRIBUTES := ( .. );
   INITIAL_THREAD_PRIORITY : constant PRIORITY := ...;


   -- for each cyclic and sporadic object  with ATC
   CONTROLLER_THREAD : array(SYSTEM_WIDE_MODE) of
             constant THREAD_ATTRIBUTES := ( .. );
   INITIAL_CONTROLLER_THREAD_PRIORITY : constant
             PRIORITY := ...;

   -- for each cyclic object
   PERIOD : array(SYSTEM_WIDE_MODE) of constant
             TIME_SPAN := ( .. );
   OFFSET : array(SYSTEM_WIDE_MODE) of constant
             TIME_SPAN := ( .. );

  -- for each sporadic object
  MAT   : array(SYSTEM_WIDE_MODE) of constant
           TIME_SPAN := ( .. );
  MAF   : array(SYSTEM_WIDE_MODE) of constant
           FREQUENCY := ( .. );
  START : array(SYSTEM_WIDE_MODE) of constant
           OPERATION_ATTRIBUTES:= ...;
end NAME_RTATT;
```

6 Mapping of Passive and Active Objects

In this chapter the mappings of *passive*, and *active* objects are considered in detail.

6.1. Passive Terminal Objects

The mapping of a *passive* terminal object <Name> to Ada 83 or Ada 95 is simply the following:

```
with <REQUIRED_INTERFACE>;
package <NAME> is

   -- for each provided type declared as private:
   type <TYPE_NAME> is private;

   -- for each  provided non-private type declared
   type <TYPE_NAME> is <TYPE_BODY>;

   -- for each provided constant declared
   <CONSTANT_NAME> : constant  <TYPE_NAME>
                   := <CONSTANT_VALUE>;

   -- for each provided operation
   procedure <OP_NAME>(<PARAMETER_PART>); -- or
   function <OP_NAME>(<PARAMETER_PART>) return <TYPE_NAME>;
   -- information on budget time and
   -- worst case execution time
```

```
    -- for each operation set
    package <SET_NAME> is
      -- for each operation in the set
      procedure <OP_NAME>(<PARAMETER_PART>); -- or
      function <OP_NAME>(<PARAMETER_PART>) return <TYPE_NAME>;
      -- information on budget time and
      -- worst case execution time
    end <SET_NAME>;

    -- for each exception
    <EXCEPTION_NAME> : exception;
    -- indication of which operations can
    -- raise the exception
  private
    -- private types with bodies declarations
    -- for each private type
    type <TYPE_NAME> is <TYPE_BODY>;

  end <NAME>;
```

For the internals, a package body:

```
  package body <NAME> is

    -- internal types, data, operations etc

    -- for each provided and internal operation
    procedure <OP_NAME>(<PARAMETER_PART>) is separate; -- or
    function <OP_NAME>(<PARAMETER_PART>)
          return <TYPE_NAME> is separate;
    procedure OPCS_OVERRUN_OF_BUDGET
                    is separate; -- not shown

    package body <SET_NAME> is separate;
  begin
    -- code of initialisation operation
  end <NAME>;
```

For each provided and internal operation:

```
with <NAME_RTATT>;
with RTA; use RTA;
with CPU_BUDGETS; use CPU_BUDGETS;

separate(<NAME>)    -- either
procedure <OP_NAME>(<PARAMETER PART>) is
  MY_ID : BUDGET_ID;
begin
  MY_ID := SET_BUDGET(<NAME_RTATT>.
     OP_NAME(CURRENT_MODE).BUDGET);
  begin
    <OPCS_CODE>;
  exception
    -- for each exception handled in the Exception handler
    -- associated to the OPCS
    when <OBJECT_NAME>.<EXCEPTION_NAME> =>
      <EXCEPTION_CODE>;
  end;
  CANCEL(MY_ID);

exception
  when BUDGET_OVERRUN =>
     if MY_ID = LAST_OVERRUN then
       OPCS_OVERRUN_OF_BUDGET;
     else
       CANCEL(MY_ID);
       raise;
     end if;
end <OP_NAME>;

-- or
function <OP_NAME>(<PARAMETER PART>) return <TYPE_NAME> is
  MY_ID : BUDGET_ID;
begin
  MY_ID := SET_BUDGET(<NAME_RTATT>.
      OP_NAME(CURRENT_MODE).BUDGET);
  begin
    <OPCS_CODE>;
  exception
    -- for each exception handled in the Exception handler
    -- associated to the OPCS
    when <OBJECT_NAME>.<EXCEPTION_NAME> =>
      <EXCEPTION_CODE>;
  end;
  CANCEL(MY_ID);
```

```
exception
  when BUDGET_OVERRUN =>
     if MY_ID = LAST_OVERRUN then
       OPCS_OVERRUN_OF_BUDGET;
     else
       CANCEL(MY_ID);
       raise;
     end if;
end <OP_NAME>;
```

6.2. Active Terminal Objects

Even within the framework given in Chapter 5 there are several possible mappings for a terminal *active* object. Terminal *active* objects are not subject to the same time constraints as other HRT-HOOD objects; the only constraint is that when hard real-time objects call *active* asynchronous operations, any blocking time is bounded. In this section we present two mappings. The first is expressed in just Ada 83 (and the extensions we have laid out in Chapter 4), it uses a protected task as a buffer; it should be noted that by implementing the buffer as a Ada 95 protected type we have a valid Ada 95 mapping. The second uses Ada 95 protected types only and therefore does not make use of the Ada 83 rendezvous mechanism; again it should be noted that this could be implemented using protected tasks in Ada 83, in which case only a restricted form of the rendezvous would be used.

Let a terminal *active* object have the following:

- ASER for an asynchronous execution request.
- LSER for a loosely synchronised operation.
- LSER_TOER for a loosely synchronised operation with an associated timeout.
- HSER for a highly synchronised operation.
- HSER_TOER for a highly synchronised operation with an associated timeout.

6.2.1. Mapping for Ada 83

The package specification of the object is:

```
with <NAME>_RTATT;
with RTA;
with SYSTEM; use SYSTEM;
with <REQUIRED_INTERFACE>;
package <NAME> is

  -- for each provided type declared as private:
  type <TYPE_NAME> is private;

  -- for each  provided non-private type declared
  type <TYPE_NAME> is <TYPE_BODY>;

  -- for each provided constant declared
  <CONSTANT_NAME> : constant <TYPE_NAME> :=
                    <CONSTANT_VALUE>;

  -- for each provided unconstrained operation
  procedure <OP_NAME>(<PARAMETER_PART>); -- or
  function <OP_NAME>(<PARAMETER_PART>)
           return <TYPE_NAME>;
  -- information on budget time and worst case
  -- execution time for each unconstrained operation set
  package <SET_NAME> is
    -- for each operation in the set
    procedure <OP_NAME>(<PARAMETER_PART>); -- or
    function <OP_NAME>(<PARAMETER_PART>)
             return <TYPE_NAME>;
    -- information on budget time and
    -- worst case execution time
  end <SET_NAME>;

  -- for each constrained operation one task entry
  task OBCS is
    entry ASER(<PARAMETER PART>);
    entry LSER(<PARAMETER PART>);
    entry HSER(<PARAMETER PART>);
    pragma PRIORITY(RTA.DEFAULT_PRIORITY);
  end OBCS;
```

```
    -- for each asynchronous constrained operation
    -- one buffer task entry
    task BUFFER is
      pragma PROTECTED_TASK;
      pragma PRIORITY(
            <NAME>_RTATT.INITIAL_PROTECTED_PRIORITY);
      entry ASER(<PARAMETER PART>);
      entry SERVER_ASER(<PARAMETER PART>);
    end BUFFER;

    procedure ASER(<PARAMETER PART>) renames BUFFER.ASER;
    procedure LSER(<PARAMETER PART>) renames OBCS.LSER;
    procedure LSER_TOER(<PARAMETER PART>;
                        TIMEOUT_DELAY : DURATION;
                        TIMEDOUT : out BOOLEAN);
    procedure HSER(<PARAMETER PART>) renames OBCS.HSER;
    procedure HSER_TOER(<PARAMETER PART>;
                        TIMEOUT_DELAY : DURATION;
                        TIMEDOUT : out BOOLEAN);

    -- for each exception
    <EXCEPTION_NAME> : exception;

  end <NAME>;
```

The body is:

```
  package body <NAME> is

    -- internal types, data, operations etc

    procedure OP_NAME(<PARAMETER PART>) is separate;
    procedure OPCS_ASER(<PARAMETER PART>) is separate;

    -- the following are not shown
    procedure OPCS_LSER(<PARAMETER PART>) is separate;
    procedure OPCS_HSER(<PARAMETER PART>) is separate;
    function OPCS_ASER_FAC(<PARAMETER PART>)
                           return BOOLEAN is separate;
    function OPCS_LSER_FAC(<PARAMETER PART>)
                           return BOOLEAN is separate;
    function OPCS_HSER_FAC(<PARAMETER PART>)
                           return BOOLEAN is separate;
```

```
task ASER_SERVER;


task body BUFFER is
  -- objects including buffer
begin
  loop
    select
      accept ASER(<PARAMETER PART>) do
        -- place data in buffer, over write buffer when
        -- full to avoid blocking
      end ASER;
    or
      when BUFFER_NOT_EMPTY =>
        accept SERVER_ASER (<PARAMETER PART>)  do
          -- return data from buffer
        end SERVER_ASER;
    or
      terminate;
    end select;
  end loop;
end BUFFER;



task body ASER_SERVER is
  -- local variable to store parameters
begin
  loop
    BUFFER.SERVER_ASER(<PARAMETER PART>);
    OBCS.ASER(<PARAMETER PART>);
  end loop;
end ASER_SERVER;
```

```
task body OBCS is
begin
  loop
    select
      when OPCS_ASER_FAC =>
        accept ASER(<PARAMETER PART>) do
          OPCS_ASER(<PARAMETER PART>);
        end ASER;
    or
      when OPCS_LSER_FAC =>
        accept LSER(<PARAMETER PART>) do
          -- copy params
        end LSER;
        OPCS_LSER(<PARAMETER PART>);
    or
      when OPCS_HSER_FAC =>
        accept HSER(<PARAMETER PART>) do
          OPCS_HSER(<PARAMETER PART>);
        end HSER;
    or
      terminate;
    end select;
  end loop;
end OBCS;

procedure LSER_TOER(<PARAMETER PART>;
                    TIMEOUT_DELAY : DURATION;
                    TIMEDOUT : out BOOLEAN) is
begin
  select
    OBCS.LSER(<PARAMETER PART>);
    TIMEDOUT := FALSE;
  or
    delay <TIMEOUT_DELAY>;
    TIMEDOUT := TRUE;
  end select;
end LSER_TOER;
```

```
   procedure HSER_TOER(<PARAMETER PART>;
                       TIMEOUT_DELAY : DURATION;
                       TIMEDOUT : out BOOLEAN) is
   begin
     select
       OBCS.HSER(<PARAMETER PART>);
       TIMEDOUT := FALSE;
     or
       delay <TIMEOUT_DELAY>;
       TIMEDOUT := TRUE;
     end select;
   end HSER_TOER;
 begin
   -- code of the initialisation operation
 end <NAME>;
```

For each unconstrained operation

```
separate(<NAME>)
procedure OP_NAME(<PARAMETER PART>) is
begin
  <OPCS_CODE>;
exception
   -- for each exception handled in the Exception
   -- handler associated to the OPCS
   when <OBJECT_NAME>.<EXCEPTION_NAME> =>
     <EXCEPTION_CODE>;
end OP_NAME;
```

For each constrained operation (for example an ASER operation):

```
separate(<NAME>)
procedure OPCS_ASER(<PARAMETER PART>) is
begin
  <OPCS_CODE>;
exception
   -- for each exception handled in the Exception
   -- handler associated to the OPCS
   when <OBJECT_NAME>.<EXCEPTION_NAME> =>
     <EXCEPTION_CODE>;
end OPCS_ASER;
```

6.2.2. Mapping for Ada 95

The main issue for the Ada 95 mapping is how the thread should wait for one of the events to occur. In Ada, a task can only wait for a single entry call and therefore it is necessary to combine the parameter parts into a single variant record (alternatively, the thread has to make two entry calls; the first returns the identity of the called method, the second gets the parameters).

Note also the use of requeue and private entries in this example.

```
with <Name>_rtatt;
with Rta;
with System; use System;
with <Required_Interface>;
package <Name> is

  -- for each provided type declared as private:
  type <Type_Name> is private;

  -- for each  provided non-private type declared
  type <Type_Name> is <Type_Body>;

  -- for each provided constant declared
  <Constant_Name> : constant  <Type_Name> :=
                                  <Constant_Value>;

  -- for each provided unconstrained operation
  procedure <Op_Name>(<Parameter_Part>); -- or
  function <Op_Name>(<Parameter_Part>) return <Type_Name>;
  -- information on budget time and
  -- worst case execution time

  -- for each unconstrained operation set
  package <Set_Name> is
    -- for each operation in the set
    procedure <Op_Name>(<Parameter_Part>); -- or
    function <Op_Name>(<Parameter_Part>)
                                  return <Type_Name>;
    -- information on budget time
    -- and worst case execution time
  end <Set_Name>;
```

```
  -- for each constrained operation,
  --  one protected type entry/procedure
  protected Obcs is
    pragma Priority(
          <Name>_rtatt.Initial_Ceiling_Priority);
    procedure Aser(<Parameter Part>);
    entry Lser(<Parameter Part>);
    entry Hser(<Parameter Part>);
    entry Get_Operation(<Variant_Parameter Part>);
    entry Set_Hser_Done(<Parameter Part>);
    entry Set_Lser_Done(<Parameter Part>);

  private
    entry WAIT_HSER_DONE(<PARAMETER PART>);
    entry WAIT_LSER_DONE(<PARAMETER PART>);
    -- buffer for ASER/LSER/HSER
    ASER_OPEN : BOOLEAN := FALSE;
    LSER_OPEN : BOOLEAN := FALSE;
    HSER_OPEN : BOOLEAN := FALSE;
    ASER_DONE : BOOLEAN := FALSE;
    LSER_DONE : BOOLEAN := FALSE;
    HSER_DONE : BOOLEAN := FALSE;
  end OBCS;


  procedure Aser(<Parameter Part>) renames Obcs.Lser;
  procedure Lser(<Parameter Part>) renames Obcs.Lser;
  procedure Lser_Toer(<Parameter Part>;
     Timeout_Delay : Duration; Timedout : out Boolean);
  procedure Hser(<Parameter Part>) renames Obcs.Hser;
  procedure Hser_Toer(<Parameter Part>;
     Timeout_Delay : Duration; Timedout : out Boolean);

  -- for each exception
  <Exception_Name> : exception;

end <Name>;
```

The body of this package is:

```
package body <Name> is

  procedure Op_Name(<Parameter Part>)
                        is separate; -- not shown
  package body <Set_Name> is separate; -- not shown

  procedure Opcs_Aser(<Parameter Part>)
                        is separate; -- not shown
  procedure Opcs_Lser(<Parameter Part>)
                        is separate; -- not shown
  procedure Opcs_Hser(<Parameter Part>)
                        is separate; -- not shown

  -- Functional activation constraints
  function Opcs_Aser_Fac return Boolean
                        is separate; -- not shown
  function Opcs_Lser_Fac return Boolean
                        is separate; -- not shown
  function Opcs_Hser_Fac return Boolean
                        is separate; -- not shown

  task Thread is
     pragma Priority(
            <Name>_rtatt.Initial_Thread_Priority);
  end Thread;

  protected body Obcs is

    procedure Aser(<Parameter Part>) is
    begin
      -- copy params to buffer
      -- handle any potential buffer overflow
      Aser_Open := True;
    end Aser;

    entry Lser(<Parameter Part>) when Opcs_Lser_Fac is
    begin
      -- copy params ; change any state which will
      -- influence OPCS_LSER_FAC
      Lser_Open := True;
      requeue Wait_Lser_Done;
    end Lser;
```

```
entry HSER(<PARAMETER PART>) when OPCS_HSER_FAC is
begin
  -- copy params; change any state which will
  -- influence OPCS_HSER_FAC
  HSER_OPEN := TRUE;
  requeue WAIT_HSER_DONE;
end HSER;

entry Get_Operation(<Variant_Parameter Part>) when
  Hser_Open or Lser_Open or
  (Aser_Open and Opcs_Aser_Fac) is
begin
  -- copy params, set HSER_OPEN or LSER_OPEN or
  -- ASER_OPEN to false if appropriate
end Get_Operation;

procedure Set_Lser_Done(<Parameter Part>) is
begin
  -- copy params
  Lser_Done := True;

end Set_Lser_Done;

procedure Set_Hser_Done(<Parameter Part>) is
begin
  -- copy params
  Hser_Done := True;

end Set_Lser_Done;

entry Wait_Hser_Done(<Parameter Part>) when
              Hser_Done is
begin
  -- copy params
  Hser_Done := False;
end Wait_Hser_Done;

entry Wait_Lser_Done(<Parameter Part>) when
              Lser_Done is
begin
  -- copy params
  Lser_Done := False;
end Wait_Lser_Done;

end Obcs;
```

```
task body Thread is
  -- any local variables required for parameter passing
begin

  loop
    begin
      -- ideally need a multiway
      -- entry call facility here
      Obcs.Get_Operation(<Variant_Parameter Part>);
      case (<Variant_Parameter Part>) is
         when Aser =>
            Opcs_Aser(<Parameter Part>);
         when Lser =>
            Obcs.Set_Lser_Done(<Parameter Part>);
            Opcs_Lser(<Parameter Part>);
         when Hser
            Opcs_Hser(<Parameter Part>);
            Obcs.Set_Hser_Done(<Parameter Part>);
      end case;
  end loop;

end Thread;


procedure Lser_Toer(<Parameter Part>;
          Timeout_Delay :
          Duration; Timedout : out Boolean) is
begin
  select
    Obcs.Lser(<Parameter Part>);
    Timedout := False;
  or
    delay <Timeout_Delay>;
    Timedout := True;
  end select;
end Lser_Toer;
```

```
    procedure Hser_Toer(<Parameter Part>;
              Timeout_Delay :
              Duration; Timedout : out Boolean) is
    begin
      select
        Obcs.Hser(<Parameter Part>);
        Timedout := False;
      or
        delay <Timeout_Delay>;
        Timedout := True;
      end select;
    end Hser_Toer;
  end <Name>;
```

6.3. Class and Instance Terminal Objects

Class objects are mapped to Ada generic packages. The specification of the package is the same as that for the corresponding object. The generic part contains the formal generic parameters. For example the following represents a *passive class* object.

```
with <REQUIRED_INTERFACE>;
generic
  -- appropriate parameter specification
  type PARAM1 is private;
  type PARAM2 is (<>);
  type PARAM3 is range <>;
  type PARAM4 is DIGIT <>;
  type PARAM5 is delta <>;

package <NAME> is

  -- for each provided type declared as private:
  type <TYPE_NAME> is private;

  -- for each  provided non-private type declared
  type <TYPE_NAME> is <TYPE_BODY>;

  -- for each provided constant declared
  <CONSTANT_NAME> : constant  <TYPE_NAME> :=
                               <CONSTANT_VALUE>;
```

```
    -- for each provided operation
    procedure <OP_NAME> (<PARAMETER_PART>); -- or
    function <OP_NAME> (<PARAMETER_PART>) return <TYPE_NAME>;
    -- information on budget time and worst case
    -- execution time

    -- for each operation set
    package <SET_NAME> is
      -- for each operation in the set
      procedure <OP_NAME> (<PARAMETER_PART>); -- or
      function <OP_NAME> (<PARAMETER_PART>) return
                                     <TYPE_NAME>;
      -- information on budget time and worst case
      -- execution time
    end <SET_NAME>;

    -- for each exception
    <EXCEPTION_NAME> : exception;
    -- indication of which operations can raise the exception
  private
    -- private types with bodies declarations
    -- for each private type
    type <TYPE_NAME> is <TYPE_BODY>;

  end <NAME>;
```

Instance objects are instantiations of the generic package.

```
  with <REQUIRED_INTERFACE>;
  package <INSTANCE_NAME> is new
          <CLASS_NAME>(ACTUAL_PARAMETERS);
```

7 Mapping Protected, Cyclic and Sporadic Objects

7.1. Protected Terminal Objects

A *protected* object controls access to shared data, or non preemptable devices. Generally the data is accessed under mutual exclusion, however, the constrained operations may not immediately require mutual exclusion. Consequently it is the definition of the operations in the ODS which define which operations require exclusive access to the data. It is assumed that a constrained operation, if itself is not "protected", will call an internal operation which is protected.

In this section we consider three mappings for a *protected* object. The first uses an Ada 83 compilation structure and assumes that mutually exclusive access has been guaranteed by the priority ordering of (or the use of offsets between) those objects which access the *protected* object. The second mapping uses Ada 83 and a protected task; and the third uses Ada 95.

Let the *protected* object <NAME> have the following:

- OP_NAME for a non-constrained operation.
- PSER for a synchronous protected operation which requires mutual exclusion immediately
- PAER for an asynchronous protected operation which requires mutual exclusion immediately
- PSERD for a synchronous protected operation which does not require mutual exclusion immediately
- PAERD for an asynchronous protected operation which does not require mutual exclusion immediately
- FPSER for a mutual exclusion operation with a functional activation constraint.
- FPSER_TOER for an operation which may be blocked due to a functional activation constraint, and which has a timeout associated with its acceptance.

- INTERNAL for an internal operation which requires mutual exclusion.

In designing the mapping for a single/multiprocessor we assume that the time taken to execute the constrained operations is small, and hence a PAER operation will be identical to a PSER. Issues of the mapping to a distributed environment are deferred until Chapter 8.

7.1.1. Ada 83 Mapping with Priorities or Offsets Guaranteeing Mutual Exclusion

A mapping similar to *passive* objects may be used for *protected* objects in the case where

- the priorities of the "using" objects have been set to ensure no conflict of access (on a single processor) or
- time offsets between "using" objects guarantee no conflict (on a single processor, or a multiprocessor).

```
with <REQUIRED_INTERFACE>;
package <NAME> is

  -- for each provided type declared as private:
  type <TYPE_NAME> is private;

  -- for each  provided non-private type declared
  type <TYPE_NAME> is <TYPE_BODY>;

  -- for each provided constant declared
  <CONSTANT_NAME> : constant  <TYPE_NAME> :=
                              <CONSTANT_VALUE>;

  -- for each provided operation (constrained or
  -- unconstrained)

  procedure <OP_NAME> (<PARAMETER_PART>); -- or
  function <OP_NAME> (<PARAMETER_PART>) return <TYPE_NAME>;

  -- information on budget time and worst case
  -- execution time
```

```
    -- for each operation set
    package <SET_NAME> is
      -- for each operation in the set
      procedure <OP_NAME> (<PARAMETER_PART>); -- or
      function <OP_NAME> (<PARAMETER_PART>)
                                 return <TYPE_NAME>;
      -- information on budget time and worst case
      -- execution time
    end <SET_NAME>;

    -- for each exception
    <EXCEPTION_NAME> : exception;
     -- indication of which operations can
     -- raise the exception
  private
    -- private types with bodies declarations
    -- for each private type
    type <TYPE_NAME> is <TYPE_BODY>;
  end <NAME>;
```

The body of the package is the same in structure to that of a *passive* object (see Chapter 6).

7.1.2. Ada 83 Mapping for a Protected Terminal Object on a Single/multi Processor

The specification of the package giving the provided operations is:

```
  with <NAME>_RTATT;
  with SYSTEM; use SYSTEM;
  with <REQUIRED_INTERFACE>;
  package <NAME> is

    -- for each non-constrained operation
    procedure OP_NAME(<PARAMETER PART>);

    task OBCS is
      pragma CEILING_PRIORITY (
             <NAME>_RTATT.INITIAL_PROTECTED_PRIORITY);
      pragma PROTECTED_TASK;
      -- for each constrained operation
      entry PSER(<PARAMETER PART>);
      entry PAER(<PARAMETER PART>);
      entry FPSER(<PARAMETER PART>);
      entry INTERNAL(<PARAMETER PART>);
    end OBCS;
```

```
  procedure PSER(<PARAMETER PART>) renames OBCS.PSER;
  procedure PAER(<PARAMETER PART>) renames OBCS.PAER;
  procedure PSERD(<PARAMETER PART>);
  procedure PAERD(<PARAMETER PART>);
  procedure FPSER(<PARAMETER PART>) renames OBCS.FPSER;
  procedure FPSER_TOER(<PARAMETER PART>;
                        TIMEOUT_DELAY : DURATION;
                        TIMEDOUT : out BOOLEAN);

  -- for each exception
  <EXCEPTION_NAME> : exception;
  -- indication of which operations can
  -- raise the exception
end <NAME>;
```

The body of this object is given below.

```
with CPU_BUDGETS; use CPU_BUDGETS;
package body <NAME> is

  -- internal types, data, exceptions etc

  procedure OP_NAME(<PARAMETER PART>) is separate;
           -- not shown, same as passive
  procedure OPCS_PSER(<PARAMETER PART>) is separate;
           -- not shown, same as above
  procedure OPCS_PAER(<PARAMETER PART>) is separate;
           -- not shown, same as above
  procedure OPCS_FPSER(<PARAMETER PART>) is separate;

           -- not shown, same as OP_Name
  function OPCS_FPSER_FAC return BOOLEAN is separate;
           -- not shown, application specific
  procedure OPCS_INTERNAL(<PARAMETER PART>) is
                                             separate;
           -- not shown, same as passive
  procedure OPCS_OVERRUN_OF_BUDGET is separate;
           -- not shown application specific
```

```
task body OBCS is
  -- no local variables
begin
  loop
    select
      when OPCS_FPSER_FAC =>
        accept FPSER(<PARAMETER PART>) do
          OPCS_FPSER(<PARAMETER PART>);
        end FPSER;
    or
      accept PSER(<PARAMETER PART>) do
          OPCS_PSER(<PARAMETER PART>);
      end PSER;
    or
      accept PAER(<PARAMETER PART>) do
          OPCS_PSER(<PARAMETER PART>);
      end PAER;
    or
      accept INTERNAL (<PARAMETER PART>) do
          OPCS_INTERNAL(<PARAMETER PART>);
      end INTERNAL;
    or
      terminate;
    end select;
  end loop;
end OBCS;

procedure FPSER_TOER(<PARAMETER PART>;
                     TIMEOUT_DELAY : DURATION;
                     TIMEDOUT : out BOOLEAN) is
begin
  select
    OBCS.FPSER(<PARAMETER PART>);
    TIMEDOUT := FALSE;
  or
    delay <TIMEOUT_DELAY>;
    TIMEDOUT := TRUE;
  end select;
end FPSER_TOER;

procedure PSERD is
begin
  -- OPCS code
  -- including a call to OBCS.INTERNAL
end;
```

```
    procedure PAERD  is
    begin
      -- OPCS code, including a call to OBCS.INTERNAL
    end;
  begin
    -- code of the initialisation operation
  end <NAME>;
```

7.1.3. Ada 95 Mapping for a Protected Terminal Object on a Single/multi Processor

There are two possible mappings for an Ada 95 *protected* object. The first is similar to above; the protected data is held global to the package and only accessed through the OBCS protected type. The second mapping encapsulates the protected data in the protected types and the OPCS routines accessing the data are generated inline instead of separate. We illustrate the latter mapping here.

The specification of the package defining the provided operations etc is:

```
with <Name>_rtatt;
with System; use System;
with Required_Objects;
package <Name> is

  -- for each non-constrained operation
  procedure Op_Name(<Parameter Part>);

  protected Obcs is
    pragma Priority (
           <Name>_rtatt.Initial_Protected_Priority);
    -- for each constrained operation
    procedure Pser(<Parameter Part>);
    procedure Paer(<Parameter Part>);
    entry Fpser(<Parameter Part>);
    procedure Internal(<Parameter Part>);
  private
    -- data to be accessed under mutual exclusion
  end Obcs;
```

```
    procedure Pser(<Parameter Part>) renames Obcs.Pser;
    procedure Paer(<Parameter Part>) renames Obcs.Paer;
    procedure Pserd(<Parameter Part>);
    procedure Paerd(<Parameter Part>);
    procedure Fpser(<Parameter Part>) renames
                                               Obcs.Fpser;
    procedure Fpser_Toer(<Parameter Part>;
                         Timeout_Delay : Duration;
                         Timedout : out Boolean);

    -- for each exception
    <Exception_Name> : exception;
    -- indication of which operations can
    -- raise the exception
  end <Name>;
```

The package body is:

```
  with Cpu_Budgets; use Cpu_Budgets;
  package body <Name> is

    -- internal types, data, exceptions etc

    procedure Op_Name(<Parameter Part>) is separate;
              -- not shown, same as passive
    procedure Opcs_Overrun_Of_Budget is separate;
              -- not shown application specific

    protected body Obcs is

      entry Fpser(<Parameter Part>) when
                  Inline_Opcs_Fpser_Fac is
        -- OPCS code generated inline;
      end Fpser;

      procedure Pser(<Parameter Part>) is
        -- OPCS code generated inline;
      end Pser;

      procedure Paer(<Parameter Part>) is
        -- OPCS code generated inline;
      end Paer;
```

```
      procedure Internal(<Parameter Part>) is
          -- OPCS code generated inline;
      end Internal;
    end Obcs;

    procedure Fpser_Toer(<Parameter Part>;
                         Timeout_Delay : Duration;
                         Timedout : out Boolean) is
    begin
      select
        Obcs.Fpser(<Parameter Part>);
        Timedout := False;
      or
        delay <Timeout_Delay>;
        Timedout := True;
      end select;
     end Fpser_Toer;

     procedure Pserd is
     begin
      -- OPCS code, including a call to OBCS.INTERNAL
     end;

     procedure Paerd  is
     begin
      -- OPCS code, including a call to OBCS.INTERNAL
     end;
  begin
     -- code of the initialisation operation
  end <Name>;
```

7.2. Cyclic Terminal Objects

There are several issues associated with the mapping of a *cyclic* object. The main one concerns the handling of asynchronous transfer of control (ATC) events. However many systems have no need for ATC; consequently we first consider an Ada 83 and Ada 95 mapping with no ATC. We then consider how to map full cyclic objects with ATC to Ada 83 and then Ada 95.

7.2.1. Cyclic Objects with No ATC Requirements

The mapping is very similar in Ada 83 or Ada 95.

```
package <NAME> is
  -- comments on real-time attributes
end <NAME>;
```

The package body is:

```
with REQUIRED_INTERFACE; use REQUIRED_INTERFACE;
with MONOTONIC; use MONOTONIC;
with <NAME>_RTATT; use <NAME>_RTATT;
with SYSTEM; use SYSTEM;
with RTA; use RTA;
package body <NAME> is

  procedure OPCS_PERIODIC_CODE is separate;
  -- not shown, similar to a passive operation
  procedure OPCS_OVERRUN_OF_BUDGET is separate;
  -- not shown

  procedure OPCS_INITIALISE(<PARAMETER PART>) is
                             separate; -- not shown

  task THREAD is
    pragma PRIORITY(
           <NAME>_RTATT.INITIAL_THREAD_PRIORITY);
  end THREAD;

  task body THREAD is
    T: TIME;
  begin
    -- any initialisation code
    OPCS_INITIALISE(<PARAMETER PART>);

    -- if the THREAD has an offset
    T := <NAME>_RTATT.OFFSET(CURRENT_MODE) +
         SYSTEM_START_UP_TIME;
```

```
      -- otherwise
      T:= CLOCK;
      loop
        delay until T;
        -- Ada 95 or "delay_until(T)" Ada 83
        OPCS_PERIODIC_CODE;
        T := T + <NAME>_RTATT.PERIOD(CURRENT_MODE);
      end loop;
    end THREAD;
  begin
    -- code of the initialisation operation
  end <NAME>;
```

It should be noted that this mapping does not detect deadline overrun; the assumption being that deadline overrun is prevented by monitoring the worst case execution time of tasks. In Ada 95, deadline overrun can be simply detected by using the asynchronous select statement:

```
  loop
    delay until T;
    select
      delay until (T + <NAME>_RTATT.
                          THREAD(CURRENT_MODE).DEADLINE);
      -- deadline overrun
    then abort
      OPCS_PERIODIC_CODE;
    end select;
    T := T + <NAME>_RTATT.PERIOD(CURRENT_MODE);
  end loop;
```

7.2.2. Mapping to Ada 83 with ATC

In this section we shall assume that only ASATC requests are required. Also, we do not consider unconstrained operations or exceptions as they have a similar mapping to unconstrained operations and exceptions in other objects.

The solution in Ada 83 requires three tasks. Two synchronised cyclics (which represent the local THREAD of the cyclic) and a protected task for communicating with the clients. The specification of the package is:

```
with REQUIRED_INTERFACE; use REQUIRED_INTERFACE;
with SYSTEM; use SYSTEM;
with <NAME>_RTATT; use <NAME>_RTATT;
package <NAME> is
  -- for each constrained operation one task entry
  task OBCS is
    pragma PRIORITY(
           <NAME>_RTATT.INITIAL_PROTECTED_PRIORITY);
    entry ASATC(<PARAMETER PART>);
    entry THREAD_FINISHED;
    entry CONTROLLER_WAIT(ATC : out BOOLEAN;
                          <PARAMETER PART>);
  end OBCS;

  procedure ASATC(<PARAMETER PART>) renames
                                        OBCS.ASATC;
end <NAME>;
```

The package body is:

```
with CPU_BUDGETS; use CPU_BUDGETS;
with REAL_TIME; use REAL_TIME;
with RTA; use RTA;
package body <NAME> is
  type PB is access BUDGET_ID;
  type PT is access TIME;

  task THREAD_CONTROLLER is
    pragma PRIORITY(<NAME>_RTATT.
              INITIAL_CONTROLLER_THREAD_PRIORITY);
  end THREAD_CONTROLLER;

  -- all dynamic memory access in the following mapping
  -- should be optimised to point to static objects
  task type THREAD is
    pragma PRIORITY(<NAME>_RTATT.INITIAL_THREAD_PRIORITY);
  end THREAD;
  type PTHREAD is access THREAD;
```

```
T: PT := new TIME'(SYSTEM_START_UP_TIME +
          <NAME>_RTATT.OFFSET(CURRENT_MODE));
     -- if the THREAD has an offset
     -- otherwise T.all := clock;
pragma SHARED(T);
MY_ID : PB := new BUDGET_ID ;
pragma SHARED(MY_ID);
ASATC_OCCURRED, THREAD_OK,
ATC_OCCURRED : BOOLEAN := FALSE;

procedure OPCS_PERIODIC_CODE is separate;
procedure OPCS_OVERRUN_OF_BUDGET is separate;
-- not shown

procedure OPCS_ASATC(<PARAMETER PART>) is
          separate; -- not shown

task body OBCS is
begin
  loop
    select
      accept ASATC(<PARAMETER PART>) do
        ASATC_OCCURRED := TRUE;
        -- save params
      end ASATC;
    or
      accept THREAD_FINISHED do
        THREAD_OK := TRUE;
      end THREAD_FINISHED;

    or
      when THREAD_OK or ASATC_OCCURRED =>
        accept CONTROLLER_WAIT(ATC : out BOOLEAN;
                      <PARAMETER PART>) do
          ATC := ASATC_OCCURRED;
          ASATC_OCCURRED := FALSE;
          THREAD_OK := FALSE;
          -- set up return parameters
       end CONTROLLER_WAIT;
    or
      terminate;
   end select;
  end loop;
end OBCS;
```

```
  -- period is the same as cyclic thread the worst case
  -- blocking time is equal to the deadline of the
  -- cyclic thread the priority is greater than the
  -- cyclic thread
  task body THREAD_CONTROLLER is
    ATHREAD : PTHREAD := new THREAD;
  begin
    loop
      OBCS.CONTROLLER_WAIT(ATC_OCCURRED, PARAMS);
      if ATC_OCCURRED then
        OPCS_ASATC(PARAMS);
        abort ATHREAD.all;
        CANCEL(MY_ID.all); -- cancel budget
        T.all := T.all +
                 <NAME>_RTATT.PERIOD(CURRENT_MODE);
            -- also set up any offset relationship
        DELAY_UNTIL (T.all);
        ATHREAD := new THREAD;
       else
        T.all := T.all + <NAME>_RTATT.PERIOD(CURRENT_MODE);
        DELAY_UNTIL (T.all);
       end if;
     end loop;
  end THREAD_CONTROLLER;

  task body THREAD is
  begin
    loop
      DELAY_UNTIL (T.all);
      OPCS_PERIODIC_CODE;
      T.all := T.all + <NAME>_RTATT.PERIOD(CURRENT_MODE);
      OBCS.THREAD_FINISHED;
    end loop;
  end THREAD;
begin
  -- code of the initialisation operation
end <NAME>;
```

```
separate(<NAME>);
procedure OPCS_PERIODIC_CODE is
begin
  MY_ID.all := SET_BUDGET( <NAME>_RTATT.
            OPCS_PERIODIC_CODE(CURRENT_MODE).BUDGET);
  <OPCS_CODE>;
  CANCEL(MY_ID.all);
exception
   when BUDGET_OVERRUN =>
    if MY_ID.all = LAST_OVERRUN then
      OPCS_OVERRUN_OF_BUDGET;
    else
      -- we are in real trouble
      raise -- some appropriate exception;
    end if;
  -- for each exception handled in the Exception
  -- handler associated to the OPCS
  when <OBJECT_NAME>.<EXCEPTION_NAME> =>
    <EXCEPTION_CODE>;
end OPCS_PERIODIC_CODE ;
```

Note that changing the OBCS task to an Ada 95 protected type gives a valid Ada 95 mapping.

7.2.3. Ada 95 Mapping for a Cyclic terminal Object <Name>

The package specification is:

```
with Required_Interface; use Required_Interface;
with System; use System;
with <Name>_rtatt; use <Name>_rtatt
package <Name> is

  -- for each constrained operation one task entry
  protected Obcs is
    pragma Priority(<Name>_rtatt.
                Initial_Protected_Priority);
    procedure Asatc(<Parameter Part>);
    entry Get_Asatc(<Parameter Part>);
  private
     Asatc_Occurred : Boolean := False;
  end Obcs;

  procedure Asatc(<Parameter Part>) renames Obcs.Asatc;
end <Name>;
```

The package body is:

```
with Ada.Real_Time.Cpu_Budgets;
use Ada.Real_Time.Cpu_Budgets;
with Ada.Real_Time; use Ada.Real_Time;
with Rta; use Rta;
package body <Name> is

  procedure Opcs_Periodic_Code is separate;
  procedure Opcs_Overrun_Of_Budget is separate;
  -- not shown

  procedure Opcs_Asatc(<Parameter Part>) is
                              separate; -- not shown

  task Thread is
    pragma Priority(<Name>_rtatt.
                  Initial_Thread_Priority);
  end Thread;

  protected body OBCS is
    procedure ASATC(<PARAMETER PART>) is
    begin
      -- one only, overwrites copy params
      ASATC_OCCURRED := TRUE;
    end ASATC;

    entry Get_Asatc(<Parameter Part>) when
                          Asatc_Occurred is
    begin
      -- copy params
      Asatc_Occurred := False;
    end Get_Asatc;
  end Obcs;
```

```
  task body Thread is
    T : Time := System_Start_Up_Time +
                <Name>_rtatt.Offset(Current_Mode);
          -- if the THREAD has an offset
          -- otherwise T := clock;
  begin
    loop
      delay until (T);
      select
        Obcs.Getatc(<Parameter Part>);
        Opcs_Asatc(<Parameter Part>);
      then abort
        Opcs_Periodic_Code;
      end select;
      T := T + <Name>_rtatt.Period(Current_Mode);
    end loop;
  end Thread;
begin
  -- code of the initialisation operation
end <Name>;

separate(<Name>);
procedure Opcs_Periodic_Code is
begin
  My_Id := Set_Budget(<Name>_rtatt.
           Opcs_Periodic_Code(Current_Mode).Budget);
  <Opcs_Code>;
  Cancel(My_Id);
exception
   when Budget_Overrun =>
    if My_Id = Last_Overrun then
      Opcs_Overrun_Of_Budget;
    else
      -- we are in real trouble
      raise -- some appropriate exception;
    end if;
  -- for each exception handled in the Exception
  -- handler associated to the OPCS
  when <Object_Name>.<Exception_Name> =>
    <Exception_Code>;
end Opcs_Periodic_Code;
```

Precedence Constraints

There is not direct mapping support for precedence constrained objects. It is assumes that the decomposition of the parent *cyclic* object will provide *protected* objects to implement the precedence constraints. The precedence constraint real-time attribute is purely to aid the schedulability analysis.

7.3. Sporadic Terminal Objects

The following issues must be considered when mapping a *sporadic* object to Ada:

1) How to enforce the minimum inter-arrival gap.
2) How to enforce any maximum arrival frequency.
3) How to enforce a maximum arrival frequency with a minimum inter-arrival gap.
4) How to implement ATC

We assume that the object to be mapped has

- an asynchronous operation called START which invokes the sporadic thread.
- an asynchronous asynchronous transfer of control request called ASATC.
- there are no offsets†.

Initially we will assume no ATC and a minimum inter-arrival gap. We will then consider no ATC and a maximum arrival frequency, followed by a maximum arrival frequency with a minimum inter-arrival gap. Finally we will consider ATC and a minimum inter-arrival gap and a maximum arrival frequency. For the purpose of our discussions we will assume that *sporadic* objects are released by *periodic* and other *sporadic* objects. Where *sporadic* objects are released by interrupts, the mappings become more complex as the devices themselves have to be manipulated to prohibit further interrupts (see Burns and Wellings[33] for a discussion on analysising and implementing sporadic tasks in Ada 95.)

7.3.1. Enforcing a Minimum Inter-arrival Gap

Within this approach the main issue to be considered is the action to be taken if the start events come in too quickly. We choose to inform the sporadic server of the overrun rather than inform the client which caused the overrun. A different mapping would enable the client to be informed.

---

† To implement offsets it is necessary for the sporadic object to be passed a parameter of type `REAL_TIME.TIME` which indicates the time at which it should next run. The sporadic task then issues a delay until this time.

```
with REQUIRED_INTERFACE; use REQUIRED_INTERFACE;
with SYSTEM; use SYSTEM;
with <NAME>_RTATT; use <NAME>_RTATT;
with REAL_TIME; use REAL_TIME;
package <NAME> is
  -- comments on real-time attributes

  -- for the task operation one task entry
  task OBCS is
    pragma PRIORITY (<NAME>_RTATT.
                      INITIAL_PROTECTED_PRIORITY);
    pragma PROTECTED_TASK;
    entry START(<PARAMETER PART>);
    entry WAIT_START(<PARAMETER PART>;
        OVERRUN: out BOOLEAN; STARTED : out TIME);
  end OBCS;

  procedure START(<PARAMETER PART>) renames OBCS.START;
end <NAME>;
```

The package body is:

```
with RTA; use RTA;
package body <NAME> is

  -- data used only by protected task
  SOVERRUN : BOOLEAN := FALSE;
  START_OPEN : BOOLEAN := FALSE;
  START_TIME : TIME;

  procedure OPCS_SPORADIC_CODE is separate;
  -- not shown, similar to a passive operation
  procedure OPCS_OVERRUN_OF_BUDGET is separate;
  -- not shown

  task THREAD is
    pragma PRIORITY(<NAME>_RTATT.INITIAL_THREAD_PRIORITY);
  end THREAD;
```

```
task body OBCS is
begin
  loop
    select
      accept START (<PARAMETER PART>) do
        if START_OPEN then
          SOVERRUN := TRUE;
        else
          -- save params
          START_TIME := CLOCK; -- log time of event
          START_OPEN := TRUE;
        end if;
      end START;

    or
      when START_OPEN =>
        accept WAIT_START (<PARAMETER PART>;
                           OVERRUN: out BOOLEAN;
                           STARTED : out TIME) do
          -- write params
          STARTED := START_TIME;
          OVERRUN := SOVERRUN;
          SOVERRUN := FALSE;
          START_OPEN := FALSE;
        end WAIT_START;
    or
      terminate;
    end select;
  end loop;
end OBCS;

task body THREAD is
  T: TIME;
  STARTED : TIME;
begin
```

```
      T:= CLOCK;
      loop
        DELAY_UNTIL(T);
        -- Ada 83 or "delay until T" Ada 95
        OBCS.WAIT_START(<PARAMETER>, OVERRUN, STARTED);
        -- test for frequency overrun
        OPCS_SPORADIC_CODE;
        -- calculate the next delay period
        if T > STARTED then
          T := T + <NAME>_RTATT.MAT(CURRENT_MODE);
        else
          T := STARTED  +
               <NAME>_RTATT.MAT(CURRENT_MODE);
        end if;
      end loop;
    end THREAD;
  begin
    -- code of the initialisation operation
  end <NAME>;
```

Note that by changing the protected task to a protected type (and by encapsulating its "local" data) we have a valid Ada 95 mapping.

7.3.2. Enforce a Maximum Arrival Frequency

In this mapping we choose, in contrast with the approach above, to inform the client of the potential overrun.

A maximum arrival frequency means that a maximum number of events can be triggered within an given period of time. To keep track of the events in the current period we maintain a circular bounded buffer (called OUTSTANDING_EVENTS) of events. *We maintain all events, irrespective of whether the task has been invoked to handle the event, until the event is stale*; an event becomes stale when the time of its occurrence is less than the current time minus the period associated with the maximum arrival frequency (stale events are removed from the buffer). At any one time then the following variables maintain the state of the buffer:

FIRST

The position of the first event in the buffer.

NEXT

The position of the next event to be given to the sporadic thread. Note that this may be different from FIRST as the first event may have been seen by the sporadic but it is not yet stale.

LAST

The position of the last event received.

OUTSTANDING

The number of events in the current time period.

TO_SEE

The number of events still to be passed to the sporadic thread.

When ever a new event is signalled the OBCS task first removes all stale events from the buffer. If the buffer is full, then overrun has occurred. If the buffer is not full then the event is stored.

The following defines the mapping.

```
with REQUIRED_INTERFACE; use REQUIRED_INTERFACE;
with SYSTEM; use SYSTEM;
with <NAME>_RTATT; use <NAME>_RTATT;
with REAL_TIME; use ONOTONICEL_TIME
package <NAME> is
  -- comments on real-time attributes

  -- for the task operation one task entry
  task OBCS is
    pragma PRIORITY (<NAME>_RTATT.
                       INITIAL_PROTECTED_PRIORITY);
    pragma PROTECTED_TASK;
    entry START(<PARAMETER PART>; OVERRUN: out BOOLEAN);
    entry WAIT_START(<PARAMETER PART>);
  end OBCS;

  procedure START(<PARAMETER PART>;
            OVERRUN: out BOOLEAN) renames OBCS.START;
 end <NAME>;
```

The package body is:

```
with RTA; use RTA;
package body <NAME> is

   subtype INVOCATION_RANGE is INTEGER range 0 ..
           <NAME>_RTATT.MAF(CURRENT_MODE).NUMBER - 1;
   subtype EVENTS is INTEGER range 0 ..
           <NAME>_RTATT.MAF(CURRENT_MODE).NUMBER;
   type EVENT_PARAMS is
      record
       <PARAMS>;
       OCCURRED : TIME;
      end record;

   -- the following data is local to the protected
   -- task and in a Ada 95 mapping would be encapsulated
   -- by the protected type

   OUTSTANDING_EVENTS : array (INVOCATION_RANGE) of
      EVENT_PARAMS := (others => (OCCURRED => CLOCK));
   TO_SEE : EVENTS := 0;
   OUTSTANDING : EVENTS  := 0;
   FIRST : INVOCATION_RANGE  := 0;
   LAST : INVOCATION_RANGE  := 0;
   NEXT : INVOCATION_RANGE  := 0;
   OLD : TIME;

   procedure OPCS_SPORADIC_CODE is separate;
   -- not shown, similar to a passive operation
   procedure OPCS_OVERRUN_OF_BUDGET is separate;
   -- not shown

   task THREAD is
     pragma PRIORITY(<NAME>_RTATT.INITIAL_THREAD_PRIORITY);
   end THREAD;
```

```
task body OBCS is
begin
  loop
  select
    accept START (<PARAMETER PART>;
                  OVERRUN: out BOOLEAN) do
      OLD := CLOCK - <NAME>_RTATT.
             MAF(CURRENT_MODE).IN_PERIOD;
      while OUTSTANDING_EVENTS(FIRST).OCCURRED <=
            OLD and OUTSTANDING /= 0 loop
        FIRST := (FIRST +1) mod
            <NAME>_RTATT.MAF(CURRENT_MODE).NUMBER;
        OUTSTANDING := OUTSTANDING - 1;
      end loop;

      if OUTSTANDING = <NAME>_RTATT.
                       MAF(CURRENT_MODE).NUMBER then
        -- we have an overrun situation
        OVERRUN := TRUE;
      else
        OUTSTANDING_EVENTS(LAST).OCCURRED := CLOCK;
        -- save params
        LAST := (LAST + 1) mod
          <NAME>_RTATT.MAF(CURRENT_MODE).NUMBER;
        TO_SEE := TO_SEE + 1;
        OUTSTANDING := OUTSTANDING + 1;
        OVERRUN := FALSE;
      end if;
    end START;

   or
    when TO_SEE > 0 =>
    accept WAIT_START (<PARAMETER PART>) do
       -- assign params
       NEXT := (NEXT + 1) mod
           <NAME>_RTATT.MAF(CURRENT_MODE).NUMBER;
       TO_SEE := TO_SEE - 1;
    end WAIT_START;
   or
    terminate;
   end select;
 end loop;
 end OBCS;
```

```
    task body THREAD is
    begin
      loop
        OBCS.WAIT_START(<PARAMETER>);
        OPCS_SPORADIC_CODE;
      end loop;
    end THREAD;
  begin
    -- code of the initialisation operation
  end <NAME>;
```

Again note, changing the protected task to an Ada 95 protected type gives a valid Ada 95 mapping.

7.3.3. Enforce a Maximum Arrival Frequency and a Minimum Inter-arrival Gap

This mapping is very similar to the one above except that we need two overrun flags: one to indicate a frequency violation and the other to indicate a minimum gap violation. The interface to the package becomes:

```
with REQUIRED_INTERFACE; use REQUIRED_INTERFACE;
with SYSTEM; use SYSTEM;
with <NAME>_RTATT; use <NAME>_RTATT;
with REAL_TIME; use REAL_TIME;
package <NAME> is
  -- comments on real-time attributes

  -- for the task operation one task entry
  task OBCS is
    pragma PRIORITY (<NAME>_RTATT.
                     INITIAL_PROTECTED_PRIORITY);
    pragma PROTECTED_TASK;
    entry START(<PARAMETER PART>; OVERRUN: out BOOLEAN;
                GAP_OVERRUN : out BOOLEAN);
    entry WAIT_START(<PARAMETER PART>);
  end OBCS;

  procedure START(<PARAMETER PART>; OVERRUN: out BOOLEAN;
            GAP_OVERRUN : out BOOLEAN) renames OBCS.START;
end <NAME>;
```

The package body simply has a new OBCS task body:

```
task body OBCS is
begin
 loop
  select
    accept START (<PARAMETER PART>; OVERRUN: out BOOLEAN;
                  GAP_OVERRUN : out BOOLEAN) do
      OLD := CLOCK - <NAME>_RTATT.
                     MAF(CURRENT_MODE).IN_PERIOD;

      while OUTSTANDING_EVENTS(FIRST).OCCURRED <=
            OLD and OUTSTANDING /= 0 loop
        FIRST := (FIRST +1) mod
           <NAME>_RTATT.MAF(CURRENT_MODE).NUMBER;
        OUTSTANDING := OUTSTANDING - 1;
      end loop;

      if OUTSTANDING = <NAME>_RTATT.
                    MAF(CURRENT_MODE).NUMBER then
        -- we have a frequency overrun situation
        OVERRUN := TRUE;
        if CLOCK - OUTSTANDING_EVENTS((LAST - 1 +
                   <NAME>_RTATT.MAF(CURRENT_MODE).
                   NUMBER) mod <NAME>_RTATT.
                   MAF(CURRENT_MODE).NUMBER).
                   OCCURRED <  <NAME>_RTATT.
                   MAT(CURRENT_MODE) then
          GAP_OVERRUN := TRUE;
        else
          GAP_OVERRUN := FALSE;
        end if;

      elsif CLOCK - OUTSTANDING_EVENTS((LAST - 1 +
                  <NAME>_RTATT.MAF(CURRENT_MODE).
                  NUMBER) mod <NAME>_RTATT.
                  MAF(CURRENT_MODE).NUMBER).
                  OCCURRED < <NAME>_RTATT.
                  MAT(CURRENT_MODE) then
        GAP_OVERRUN := TRUE;
        OVERRUN := FALSE;
```

```
          else
            OUTSTANDING_EVENTS(LAST).OCCURRED := CLOCK;
            -- save params
            LAST := (LAST + 1) mod
              <NAME>_RTATT.MAF(CURRENT_MODE).NUMBER;
            TO_SEE := TO_SEE + 1;
            OUTSTANDING := OUTSTANDING + 1;
            OVERRUN := FALSE;
            GAP_OVERRUN := FALSE;
          end if;
        end START;

      or
        when TO_SEE > 0 =>
          accept WAIT_START (<PARAMETER PART>) do
             -- assign params
             NEXT := (NEXT + 1) mod
               <NAME>_RTATT.MAF(CURRENT_MODE).NUMBER;
             TO_SEE := TO_SEE - 1;
          end WAIT_START;
      or
        terminate;
      end select;
    end loop;
  end OBCS;
```

7.3.4. Enforce a Maximum Arrival Frequency and a Minimum Inter-arrival Gap and ATC

This solution combines the above solution with that for ATC given for *cyclic* objects.

Ada 83 Implementation

```
with REQUIRED_INTERFACE; use REQUIRED_INTERFACE;
with SYSTEM; use SYSTEM;
with <NAME>_RTATT; use <NAME>_RTATT;
with REAL_TIME; use REAL_TIME;
package <NAME> is
```

```
  -- comments on real-time attributes

  -- for the task operation one task entry

  task OBCS is
    pragma PRIORITY (<NAME>_RTATT.
                      INITIAL_PROTECTED_PRIORITY);
    pragma PROTECTED_TASK;

    entry START(<PARAMETER PART>; OVERRUN: out BOOLEAN;
           GAP_OVERRUN : out BOOLEAN);
    entry WAIT_START(<PARAMETER PART>);
    entry ASATC(<PARAMETER PART>);
    entry THREAD_FINISHED;
    entry CONTROLLER_WAIT(ATC : out BOOLEAN);
  end OBCS;

  procedure START(<PARAMETER PART>; OVERRUN: out BOOLEAN;
           GAP_OVERRUN : out BOOLEAN) renames OBCS.START;
  procedure ASATC(<PARAMETER PART>) renames OBCS.ASATC;
end <NAME>;
```

The package body is:

```
with RTA; use RTA;
with CPU_BUDGETS; use CPU_BUDGETS;
package body <NAME> is

  subtype INVOCATION_RANGE is INTEGER range 0 ..
          <NAME>_RTATT.MAF(CURRENT_MODE).NUMBER - 1;
  subtype EVENTS is INTEGER range 0 ..
          <NAME>_RTATT.MAF(CURRENT_MODE).NUMBER;
  type EVENT_PARAMS is
     record
      <PARAMS>;
      OCCURRED : TIME;
     end record;

  task THREAD_CONTROLLER is
    pragma PRIORITY (<NAME>_RTATT.
          INITIAL_CONTROLLER_THREAD_PRIORITY);
  end THREAD_CONTROLLER;
```

```
task type THREAD is
  pragma PRIORITY(<NAME>_RTATT.INITIAL_THREAD_PRIORITY);
end THREAD;
type PTHREAD is access THREAD;

-- the following data is local to the protected
-- task and in a Ada 95 mapping would be encapsulated
-- by the protected type

OUTSTANDING_EVENTS : array (INVOCATION_RANGE) of
  EVENT_PARAMS := (others => (OCCURRED => CLOCK));
TO_SEE : EVENTS := 0;
OUTSTANDING : EVENTS  := 0;
FIRST : INVOCATION_RANGE  := 0;
LAST : INVOCATION_RANGE  := 0;
NEXT : INVOCATION_RANGE  := 0;
OLD : TIME;

type PB is access BUDGET_ID;
type PT is access TIME;

MY_ID : PB := new BUDGET_ID ;
pragma SHARED(MY_ID);
ASATC_OCCURRED, THREAD_OK,
ATC_OCCURRED : BOOLEAN := FALSE;

procedure OPCS_SPORADIC_CODE is separate;
-- not shown, similar to a passive operation
procedure OPCS_OVERRUN_OF_BUDGET is separate;
-- not shown

task body OBCS is
begin
```

```
loop
 select
  accept START (<PARAMETER PART>;
         OVERRUN: out BOOLEAN;
         GAP_OVERRUN : out BOOLEAN) do

    OLD := CLOCK - <NAME>_RTATT.
            MAF(CURRENT_MODE).IN_PERIOD;
    while OUTSTANDING_EVENTS(FIRST).OCCURRED <=
           OLD and OUTSTANDING /= 0 loop
      FIRST := (FIRST +1) mod
         <NAME>_RTATT.MAF(CURRENT_MODE).NUMBER;
      OUTSTANDING := OUTSTANDING - 1;
    end loop;

    if OUTSTANDING = <NAME>_RTATT.
       MAF(CURRENT_MODE).NUMBER then
       -- we have an overrun situation
      OVERRUN := TRUE;
      if CLOCK - OUTSTANDING_EVENTS((LAST - 1 +
          <NAME>_RTATT.MAF(CURRENT_MODE).NUMBER)
          mod <NAME>_RTATT.MAF(CURRENT_MODE).
          NUMBER).OCCURRED <  <NAME>_RTATT.
          MAT(CURRENT_MODE) then
         GAP_OVERRUN := TRUE;
      else
        GAP_OVERRUN := FALSE;
      end if;

    elsif CLOCK - OUTSTANDING_EVENTS((LAST - 1 +
       NAME>_RTATT.MAF(CURRENT_MODE).NUMBER) mod
       <NAME>_RTATT.MAF(CURRENT_MODE).NUMBER).
       OCCURRED < <NAME>_RTATT.
       MAT(CURRENT_MODE) then
      GAP_OVERRUN := TRUE;
      OVERRUN := FALSE;
```

```
        else
          OUTSTANDING_EVENTS(LAST).OCCURRED := CLOCK;
          -- save params
          LAST := (LAST + 1) mod
             <NAME>_RTATT.MAF(CURRENT_MODE).NUMBER;
          TO_SEE := TO_SEE + 1;
          OUTSTANDING := OUTSTANDING + 1;
          OVERRUN := FALSE;
          GAP_OVERRUN := FALSE;
        end if;
      end START;

    or
      when TO_SEE > 0 =>
        accept WAIT_START (<PARAMETER PART>) do
           -- assign params
          NEXT := (NEXT + 1) mod
             <NAME>_RTATT.MAF(CURRENT_MODE).NUMBER;
          TO_SEE := TO_SEE - 1;
        end WAIT_START;
    or
      accept ASATC(<PARAMETER PART>) do
        -- copy params
        ASATC_OCCURRED := TRUE;
      end ASATC;

    or
      accept THREAD_FINISHED do
        THREAD_OK := TRUE;
      end THREAD_FINISHED;
    or
      when THREAD_OK or ASATC_OCCURRED =>
        accept CONTROLLER_WAIT(ATC : out BOOLEAN) do
          -- copy params
          ATC := ASATC_OCCURRED;
          ASATC_OCCURRED := FALSE;
          THREAD_OK := FALSE;
        end CONTROLLER_WAIT;
    or
      terminate;
    end select;
  end loop;
end OBCS;
```

```
    -- the worst case blocking time is equal to the
    -- deadline of the cyclic thread the priority
    -- is greater than the sporadic thread the
    -- cost of task creation and the activation of the
    -- thread must be charged to this task

    task body THREAD_CONTROLLER is
      ATHREAD : PTHREAD := new THREAD;
    begin
      loop
        OBCS.CONTROLLER_WAIT(ATC_OCCURRED);
        if ATC_OCCURRED then
          abort ATHREAD.all;
          CANCEL(MY_ID.all);
          ATHREAD := new THREAD;
         end if;
       end loop;
    end THREAD_CONTROLLER;

    task body THREAD is
    begin
      loop
        OBCS.WAIT_START(<PARAMETER>);
        OPCS_SPORADIC_CODE;
        OBCS.THREAD_FINISHED;
      end loop;

    end THREAD;
  begin
    -- code of the initialisation operation
  end <NAME>;
```

Ada 95 Implementation

```
  with Required_Interface; use Required_Interface;
  with System; use System;
  with <Name>_rtatt; use <Name>_rtatt;
  with Ada.Real_Time; use Ada.Real_Time;
  package <Name> is
    -- comments on real-time attributes
```

```
  subtype Invocation_Range is Integer range 0 ..
          <Name>_rtatt.Maf(Current_Mode).Number - 1;
  subtype Events is Integer range 0 ..
          <Name>_rtatt.Maf(Current_Mode).Number;
  type Event_Params is
     record
      <Params>;
      Occurred : Time;
     end record;

  -- for the task operation one task entry
  protected Obcs is
    pragma Priority (
           <Name>_rtatt.Initial_Protected_Priority);
    procedure Start(<Parameter Part>; Overrun: out Boolean;
                    Gap_Overrun : out Boolean);
    entry Wait_Start(<Parameter Part>);
    procedure Asatc(<Parameter Part>);
    entry Wait_Asatc(<Parameter Part>);

  private
    Outstanding_Events : array (Invocation_Range) of
    Event_Params := (others => (Occurred => Clock));
    To_See : Events := 0;
    Outstanding : Events  := 0;
    First : Invocation_Range  := 0;
    Last : Invocation_Range  := 0;
    Next : Invocation_Range  := 0;
    Asatc_Occurred: Boolean := False;
    Old : Time;
  end Obcs;

  procedure Start(<Parameter Part>; Overrun: out Boolean;
            Gap_Overrun : out Boolean) renames Obcs.Start;
  procedure Asatc(<Parameter Part>) renames Obcs.Asatc;
end <Name>;
```

The package body is:

```
with Rta; use Rta;
with Ada.Real_Time.Cpu_Budgets;
use Ada.Real_Time.Cpu_Budgets;
package body <Name> is
```

```
procedure Opcs_Sporadic_Code is separate;
-- not shown, similar to a passive operation
procedure Opcs_Overrun_Of_Budget is separate;
-- not shown

task Thread_Controller is
  pragma Priority (<Name>_rtatt.
         Initial_Controller_Thread_Priority);
end Thread_Controller;

task Thread is
  pragma Priority(<Name>_rtatt.
         Initial_Thread_Priority);
end Thread;

protected body Obcs is

   procedure Start (<Parameter Part>;
       Overrun: out Boolean;
       Gap_Overrun : out Boolean) is
   begin
     Old := Clock - <Name>_rtatt.
            Maf(Current_Mode).In_Period;
     while Outstanding_Events(First).Occurred <=
          Old and Outstanding /= 0 loop
      First := (First +1) mod
          <Name>_rtatt.Maf(Current_Mode).Number;
      Outstanding := Outstanding - 1;
     end loop;

     if Outstanding = <Name>_rtatt.
                Maf(Current_Mode).Number then
       -- we have an overrun situation
       Overrun := True;
     elsif Clock - Outstanding_Events((Last - 1 +
        <Name>_rtatt.Maf(Current_Mode).Number) mod
        <Name>_rtatt.Maf(Current_Mode).Number).
        Occurred < <Name>_rtatt.Mat(Current_Mode) then
      Gap_Overrun := True;
```

```
      else
       Outstanding_Events(Last).Occurred := Clock;
       -- save params
       Last := (Last + 1) mod
           <Name>_rtatt.Maf(Current_Mode).Number;
       To_See := To_See + 1;
       Outstanding := Outstanding + 1;
       Overrun := False;
       Gap_Overrun := False;
       end if;
  end Start;

  entry Wait_Start (<Parameter Part>)
                  when To_See > 0 is
  begin
     Next := (Next + 1) mod
           <Name>_rtatt.Maf(Current_Mode).Number;
     To_See := To_See - 1;
  end Wait_Start;

  procedure Asatc(<Parameter Part>) is
  begin
     -- copy params
     Asatc_Occurred := True;
  end Asatc;

  entry Wait_Asatc(<Parameter Part>)
                 when Asatc_Occurred is
  begin
    -- copy params
    Asatc_Occurred := False;
  end Controller_Wait;

end Obcs;
```

```
    task body Thread is
    begin
      loop
        select
          Obcs.Getatc(<Parameter Part>);
          Opcs_Asatc(<Parameter Part>);
        then abort
          Obcs.Wait_Start(<Parameter Part>);
          Opcs_Sporadic_Code;
        end select;
      end loop;

    end Thread;
  begin
    -- code of the initialisation operation
  end <Name>;
```

8 Distributed Systems

In HRT-HOOD groups of objects may be allocated to separate nodes in a distributed system. Communication between nodes is achieved by notionally distributed *protected* or *sporadic* objects. In this section we therefore consider how these objects may be mapped to Ada. The general approach is to use remote procedure call technology[15, 58] A *protected* object X or *sporadic* object Y will be mapped to:

- the actual object X (or Y)
- a client stub for X (or Y)
- a server stub for X (or Y)

In general there are several communication mechanisms that could be supported across the network. To aid in the schedulability analysis of distributed systems we insists that any messages sent between hard real-time objects should be asynchronous. During configuration of the terminal objects, objects which communicate asynchronously may be grouped on different machines; those which communicate synchronously, or with pointers, will be grouped on the same machine.

For each *protected* object that must be accessed remotely we create two "stub" objects. A client stub is responsible for taking the call from the client object, packing the call parameters and call identification details into a record, and then passing this record to a communication subsystem for delivery to the remote site (this is standard remote procedure call technology which is described by Birrell and Nelson[15]). As this is an asynchronous call, the client object is released as soon as the data has been passed to the communication subsystem. At the called site, the communication subsystem informs a general server that data has been received for a remote object. The server unpacks the parameters and calls the appropriate server stub to call the *protected* operation. Figure 8.1 illustrates the approach. The situation for a *sporadic* object is similar, and is illustrated in Figure 8.2.

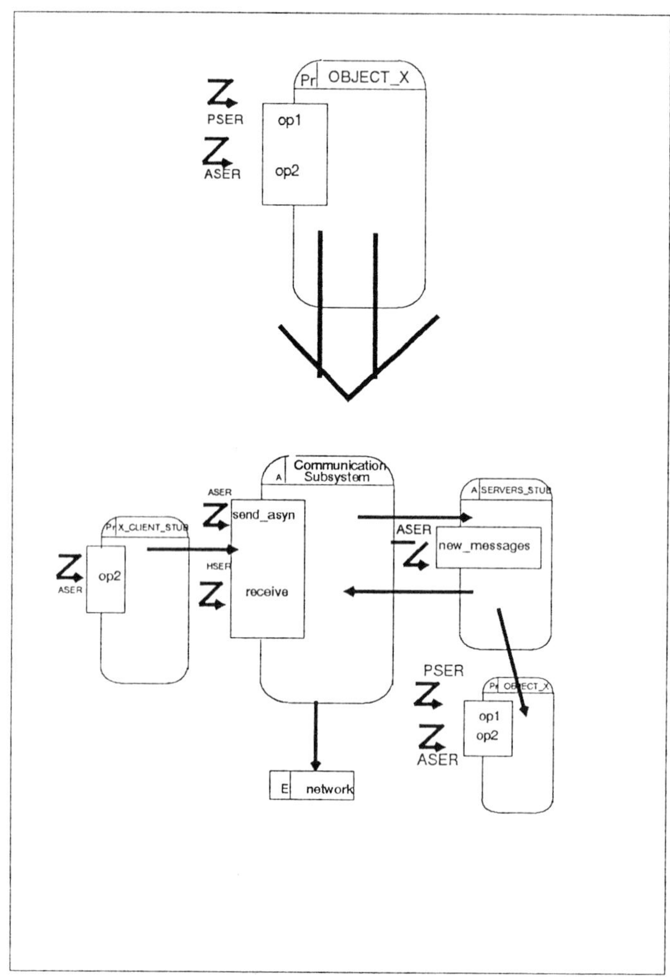

Figure 8.1 Translation of a Distributed Protected Object

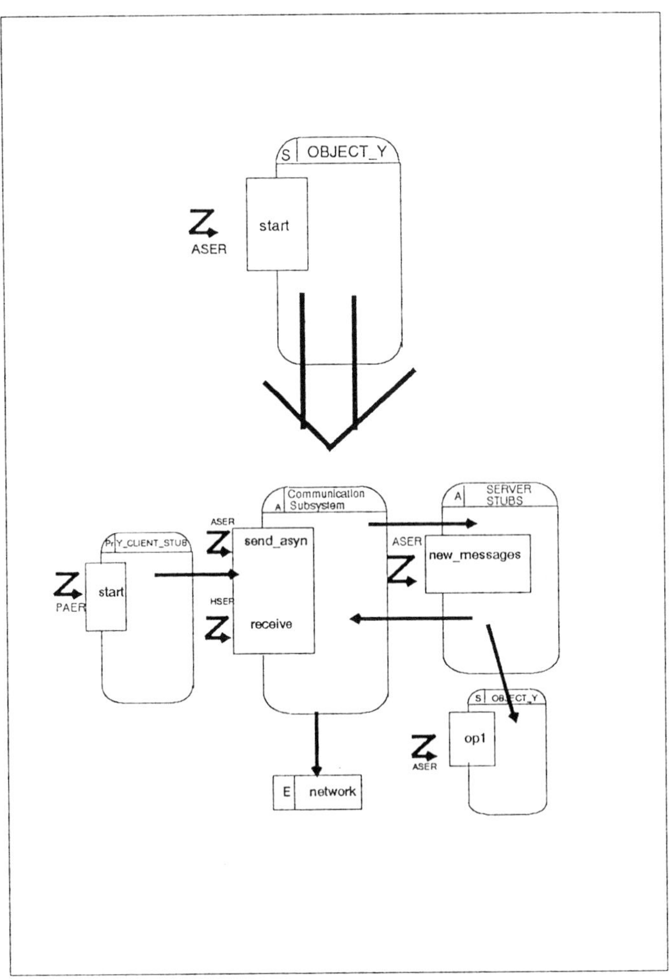

Figure 8.2 Translation of a Distributed Sporadic Object

8.1. Analysable Communication Subsystems

In this section we consider the implementation of the communication subsystem. It is not the goal of this book to dictate use of a particular communications approach but rather to indicate the issues that must be addressed if the time taken to communicate across the network is bounded. For the purpose of this discussion we consider a token passing network which only interrupts the host node when the token arrives. We assume that data destined for the host can be stored in the network buffers (maintained by the Network *environment* object). The subsystem is viewed as an *active* object with operations to send and receive messages across the network, and an operation to indicate that the token has arrived. The object is depicted in Figure 8.3.

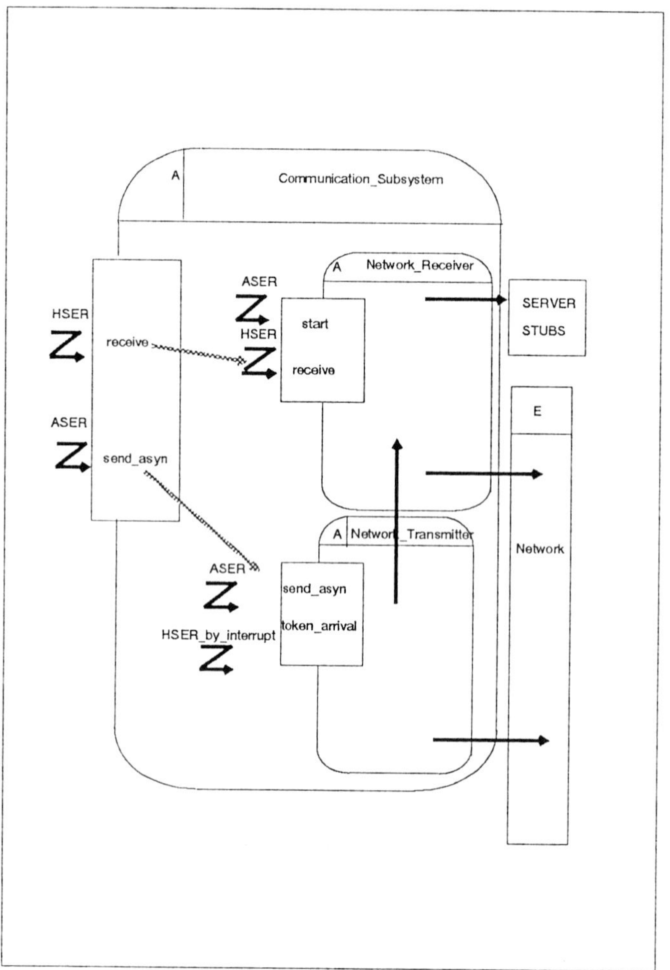

Figure 8.3 The Communication Subsystem

The communication subsystem is decomposed into two *active* objects. One which handles the transmission of messages, the other handles the reception of messages. The Network_Transmitter receives the token_arrival interrupt. When it has finished transmitting messages it invokes the Network_Receiver object to handle the messages that might have been received since it was last invoked. The transmission component of the protocol is illustrated in Figure 8.4.

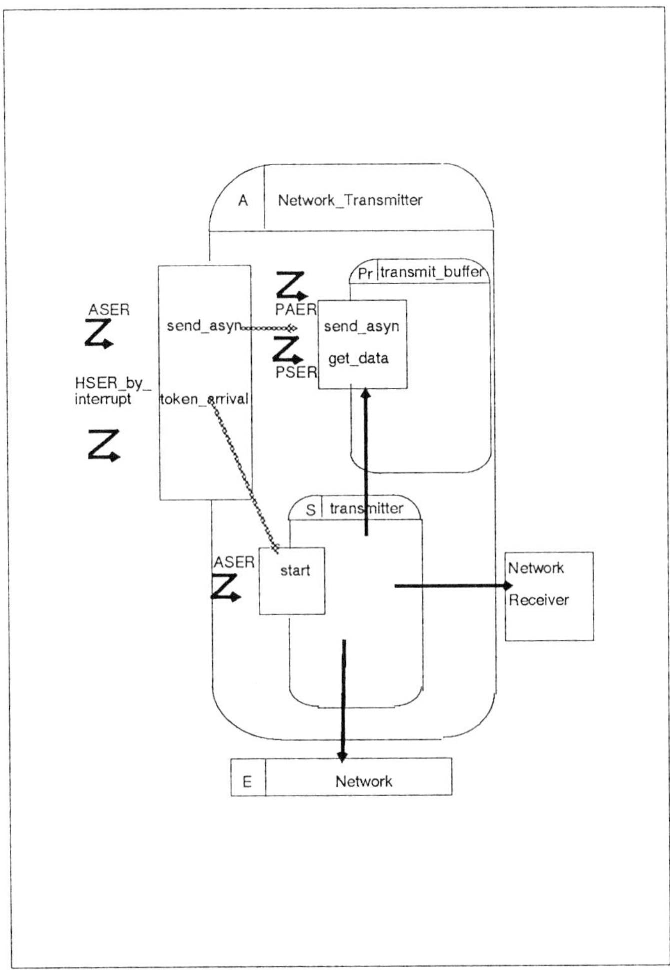

Figure 8.4: The Network Transmitter

In this model we assume that data to be sent across the network is stored in a local buffer before being copied into the network buffer. If the network buffer has appropriate characteristics, the data may be placed directly in it. On receipt of a token arrival interrupt, the transmit process copies data into the network buffer and busy-waits on the transmit (some double buffering techniques may be used to cut down on busy waiting time). Note that the transmitter process itself might also be performed by hardware. Having transmitted all the data, the Network_Transmitter passes on the token and then instructs the Network_Receiver object to check for received data. The Network_Receiver object is illustrated in Figure 8.5.

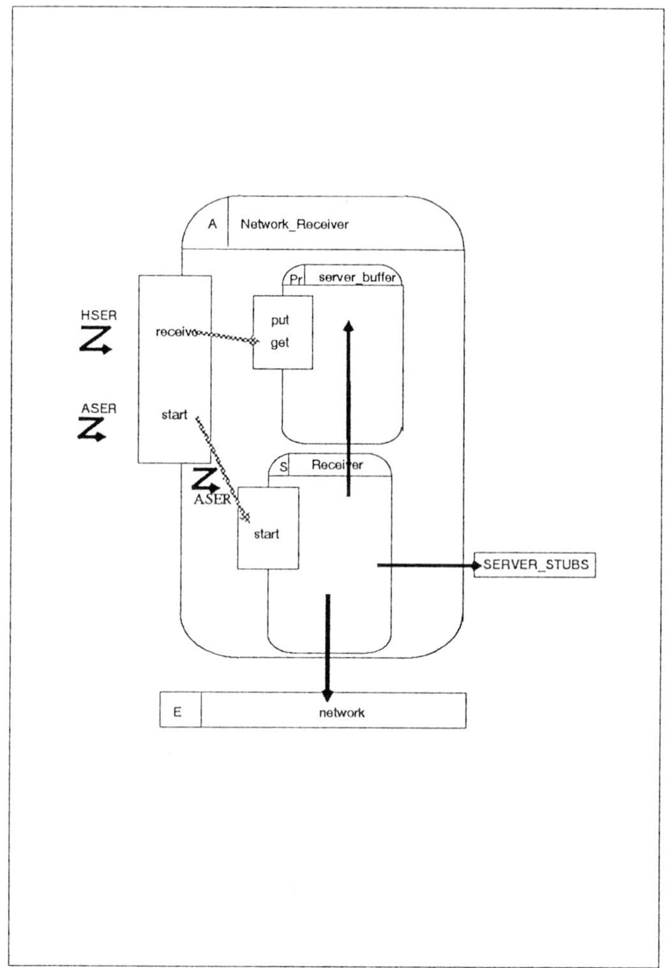

Figure 8.5: The Network Receiver

Once started, the Receiver object copies the data from the network buffer to local buffer space; it then informs the SERVER_STUBS object that new data has arrived and who are the intended recipients. The server objects will then call "receive" to get the data addressed to them.

8.1.1. Server Stubs

There is one SERVER_STUBS object for all objects on a node which can be accessed remotely. Its design is illustrated in Figure 8.6.

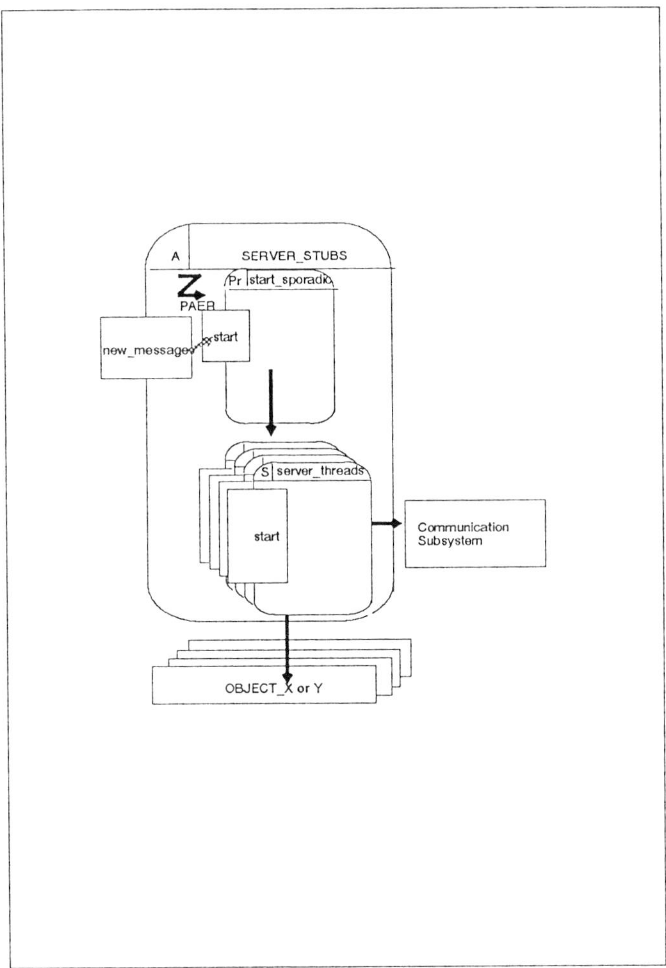

Figure 8.6: Server Stubs

On invocation of the new_messages operation, the start_sporadic object calls the start operation of the appropriate server_thread which is responsible for reclaiming the message from the network, unpacking the data (via the server stub) etc., and then calling its corresponding *protected/sporadic* object.

8.1.2. Summary

Figure 8.7 summarises the interactions between the objects of the communication subsystem.

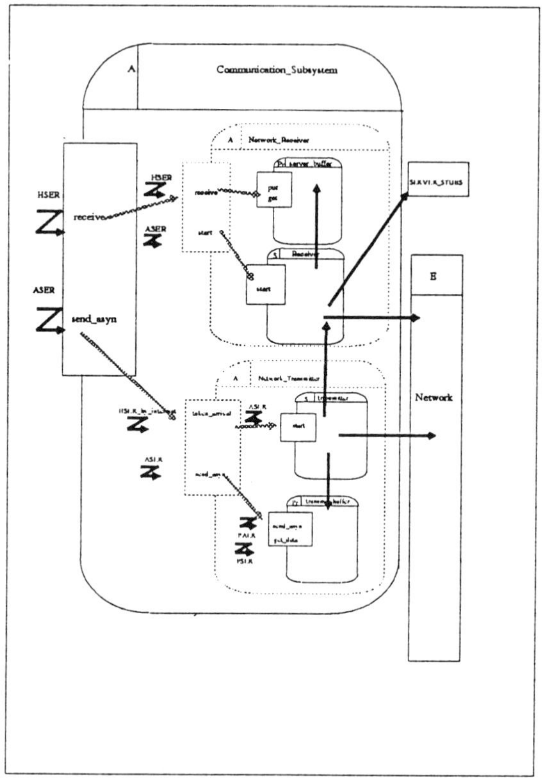

Figure 8.7: Interaction with Communication Subsystem

8.2. Mapping to Ada 95

In this section we consider some of the issues of mapping HRT-HOOD to the distribution facilities of Ada 95.

8.2.1. Ada 95 and Distributed Systems

Ada 95 considers both shared memory and distributed (non shared memory) architectures. For distributed system their support provides for a virtual-node[49] like approach, with remote subprogram call being the sole method of communication. The virtual node is called an active partition. A program consists of one or more active partitions and zero or more passive partitions. Each active partition consists of one or more library units, one of which may be designated as a main subprogram. The interface to an active partition is specified via zero or more `Remote_Call_Interface` packages. Calling subprograms declared in such a package from another active partition result in remote subprogram calls, with the parameters and results traveling via a user-provided partition communication subsystem (PCS) which interconnects the active partitions. The types declared as parameters in a remote subprogram can be declared in a package identified with a `Remote_Types` or `Pure pragma`.

As with virtual nodes, the active partition is the unit of distribution. The PCS may include application specific facilities for message passing etc. At a minimum it must support the RPC facility (see later). An active partition is presumed to execute on a set of one or more tightly-coupled logical processors with some amount of fully-shared memory. One or more active partitions may execute on a single processing unit

The library units which comprise an active partition are elaborated by an environment task devoted to the partition, and then, if present, the body of the main subprogram is executed. Upon completion of the main subprogram (if any), the environment task waits until all tasks dependent on it terminate, and then it finalises the library units which comprise the active partition.

Passive partitions consist of one or more passive library units, and represent data visible to one or more active partitions, normally implemented with a physically shared memory module. A passive library unit must be either pure or preelaborated, and has no library-level tasks, nor library-level protected types with entries. A pure library unit contains no library-level variables or access collections, and depends only on other pure packages or pure subprograms. A preelaborated library unit package has no declarations which require the execution of statements or the evaluation of a variable to perform their elaboration, and depends only on pure or preelaborated library units. Preelaborated variables must be of a type with predefined copy and finalisation. Such a package is elaborated prior to all non-preelaborated packages, in an order consistent with the dependences. Note that a preelaborated library unit can include an access type definition. A preelaborated passive library unit is called a `Shared_Passive`. There is no environment task associated with a passive partition.

Active partitions may be added to or deleted from a program during its execution, although it is not specified how this can be achieved. It is assumed that each active partition has a associated partition identifier. Attempts to communicate with an active partition which is not present (or not reachable) will raise the (new) predefined exception `Communication_Error`. Deleting an active partition is equivalent to aborting its environment task. Adding an active partition creates a new environment task which elaborates the partition and executes its main subprogram, if any. Because partitions can be added or created at any time, it is not defined when a distributed Ada program should terminate. In general, a program should not terminate if it haş active partitions which have not terminated.

The Partition Communication Subsystem

Ada has defined how distributed programs can be partitioned and what forms of remote communication must be supported. However, the language designers were keen not to over specify the language and not to prescribe a distributed run-time support system for Ada programs. They wanted to allow implementors to provide their own network communication protocols and where appropriate allow other ISO standards to be used; for example the ISO Remote Procedure Call standard. To achieve these aims, the Ada language assumes the existence of a standard implementation-provided package for handling all remote communication (the Partition Communication Subsystem, PCS). This allows compilers to generate calls to a standard interface without being concerned with the underlying implementation.

The following package defines the interface to a remote procedure (subprogram) call (RPC) support system.

```
with Ada.Streams;
package System.RPC is

   type Partition_ID is range 0 .. implementation_defined;

   Communication_Error : exception;

   type Params_Stream_Type(Initial_Size :
        Ada.Streams.Stream_Element_Count) is new
        Ada.Streams.Root_Stream_Type with private;

   procedure Read(Stream : in out Params_Stream_Type;
           Item : out Ada.Streams.Stream_Element_Array;
           Last : out Ada.Streams.Stream_Element_Offset);

   procedure Write(Stream : in out Params_Stream_Type;
           Item : in Ada.Streams.Stream_Element_Array);

   -- Synchronous call
   procedure Do_RPC(Partition   : in Partition_ID;
           Params : access Params_Stream_Type;
           Result : access Params_Stream_Type);
 -- Asynchronous call
   procedure Do_APC(Partition   : in Partition_ID;
           Params : access Params_Stream_Type);
```

```
   -- The handler for incoming RPCs
   type RPC_Receiver is access procedure(
           Params : access Params_Stream_Type;
           Result : access Params_Stream_Type);

   procedure Establish_RPC_Receiver(
           Receiver : in RPC_Receiver);

private
   ... -- not specified by the language
end System.RPC;
```

The type Partition_Id is used to identify partitions. For any library level declaration, D, D'Partition_Id yields the identifier of the partition in which the declaration was elaborated.

The exception Communication_Error is raised when an error is detected by System.RPC during a remote procedure call. Note: only one exception is specified even though many sources of errors might exist, since it is not always possible to distinguish among these errors. In particular, it is often impossible to tell the difference between a failing communication link and a failing processing element. The function Exception_Information may be used by an implementation to provide more detailed information about the cause of the exception, such as communication errors, missing partitions, etc.

An object of stream type Params_Stream_Type is used for marshalling (trans lating data into a appropriate stream-oriented form) and unmarshalling the parameters or results of a remote subprogram call, for the purposes of sending them between partitions. The object is also used to identify the particular subprogram in the called partition.

Two abstract operations are defined in a standard package called Streams to support the interface: 'Write and 'Read. 'Write marshals an object so that its representation is suitable for transmission across the network. the object once it is been transmitted.

The compiler will enforce restrictions on the Remote_Call_Interface packages, and provides stubs to be called from remote active partitions. These stubs pack up parameters, invoke the user provided communication subsystem, and process any **in** and **in out** parameters or exceptions from the acknowledgement messages when the RPC completes. Corresponding routines are generated to handle the RPC at the server site, which receive the RPC messages, invoke the normal Ada subprograms, and report back the results.

The procedure Do_RPC is invoked by the calling stub after the parameters are flattened into the message. After sending the message to the remote partition, it suspends the calling task, until a reply arrives. The procedure Do_APC acts like Do_RPC except that it returns immediately after sending the message to the remote partition. It is called whenever the Asynchronous pragma is specified for the remotely called procedure. Establish_RPC_Receiver is called immediately after elaborating an active partition, but prior to invoking the main subprogram, if any. The Receiver parameter designates an implementation-provided procedure that receives a message and calls the appropriate RCI package and subprogram.

If a package is assigned to the same partition as a Remote_Call_Interface package, then it is valid for an implementation to optimise all calls to the RCI package so that they are local calls (rather than going via the PCS — Partition Communication Subsystem).

The pragma All_Calls_Remotewhen applied to a package indicates that all calls must be routed through the PCS. This allows, for example, a distributed system to be debugged more easily on a single processor system.

8.3. Mapping Protected Objects in a Distributed Ada Environment

It is beyond the scope of this book to fully define the infrastructure required to support partitioning and distributed execution. We assume that the techniques that have been used to implement virtual nodes in an Ada 83 environment can be used,[5,50] and therefore will not discuss further an Ada 83 mapping. Here we briefly illustrate the how protected objects are mapped to the Ada 95 Distribution Annex. Consider the following object:

```
package Protected_Object is
  -- types and exceptions declared by the object

  -- operation declarations
  protected Obcs is
    procedure Op1(...);
    procedure Op2(...);
  private
    -- local data
  end ;

  procedure Op1(...) renames Obcs.Op1;
  procedure Op2(...) renames Obcs.Op2;

end Protected_Object;
```

```
package body Protected_Object is
  -- operation implementation
end Protected_Object;
```

For protected objects which are to be distributed it is necessary to introduce a layer of indirection into the mapping because Ada 95 does not allow a protected object to appear in the interface of a partition.

```
package Protected_Object.Remote_Interface is
  pragma Remote_Call_Interface;

  pragma Asynchronous(Op1);
  procedure Op1(...) ;
  procedure  Op2(...) ;
  -- not present if synchronous call are
  -- not allowed across the network

end Protected_Object;

with Protected_Object;
package body
        Protected_Object.Remote_Call_Interface is
  procedure Op1(...) is
  begin
    Protected_Object.Op1(..);
  end Op1;
  procedure Op2(...) is
  begin
    Protected_Object.Op2(..);
  end Op2;
end Protected_Object;
```

Note that the compiler will generate the client and server stubs. The compiler generated client stubs automatically call the `System.RPC` package. This interface can be mapped to the interface provided by the Communication_Subsystem describe above. When the RPC is delivered to the server threads they can call the compiler supported server stud to deliver the call to the actual object.

Part 3
Case Studies

HRT-HOOD has been designed primarily to support hard real-time space applications. However, it is intended that it should be a general purpose design method for hard and soft real-time systems. Consequently, this book presents two case studies. The first is for pedagogical purposes and details the design and implementation of a mine control system. The example that has been chosen is based on one which commonly appears in the literature[18, 19, 56, 74, 76] (and one that the authors have already investigated using PAMELA[19]); it possesses many of the characteristics which typify embedded real-time systems. It is assumed that the system will be implemented on a single processor with a simple memory mapped I/O architecture.

The second study considers the use of HRT-HOOD to design and reimplement the attitude and orbital control system of the Olympus Satellite.[27, 28] It was performed by British Aerospace in consulatation with the Authors.

9 The Mine Control System

The purpose of this case study is to exercise the HRT-HOOD design method. We draw upon the analysis of the problem that has previously been given by the authors.[19,25] Although we shall present estimations of execution times etc, these are for illustrative purposes only.

In Section 9.1 an overview of the mine control system is given, along with its main functional and non-functional requirements. Section 9.2 then produces a logical architecture for the proposed computer system. In Section 9.3, the physical architecture is developed by performing the schedulability analysis. This is followed, in Section 9.4 physical architecture.

9.1. Mine Control System Overview

The study concerns the design of software necessary to manage a simplified pump control system for a mining environment. The system is used to pump mine water, which collects in a sump at the bottom of the shaft, to the surface. A simple schematic diagram of the system is given in Figure 9.1.

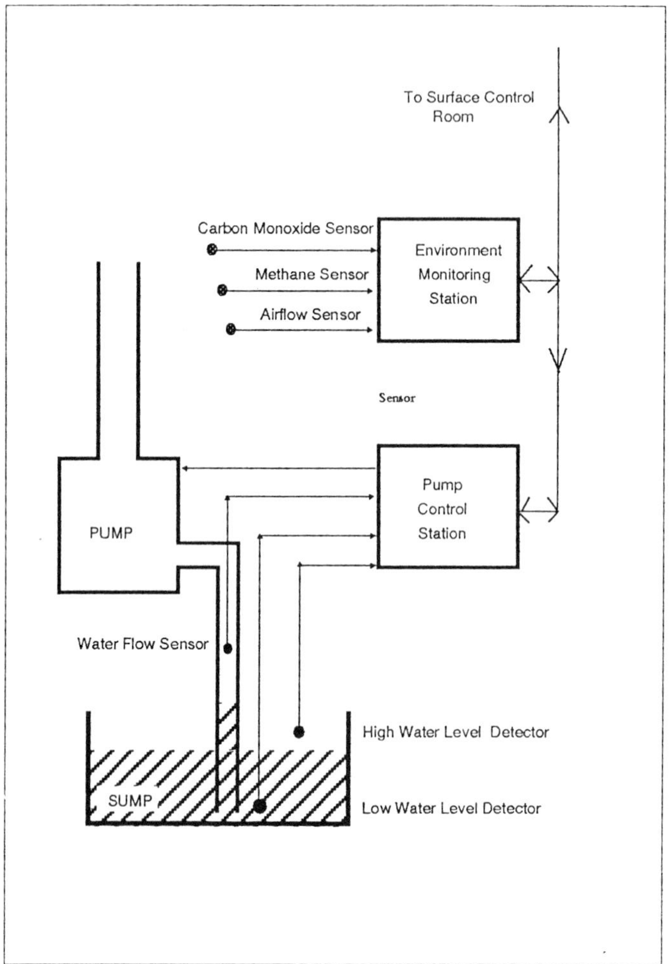

Figure 9.1: A Mine Drainage Control System

The relationship between the control system and the external devices is shown in Figure 9.2. Note that only the high and low water sensors communicate via interrupts (indicated by shaded arrows); all the other devices are either polled or directly controlled.

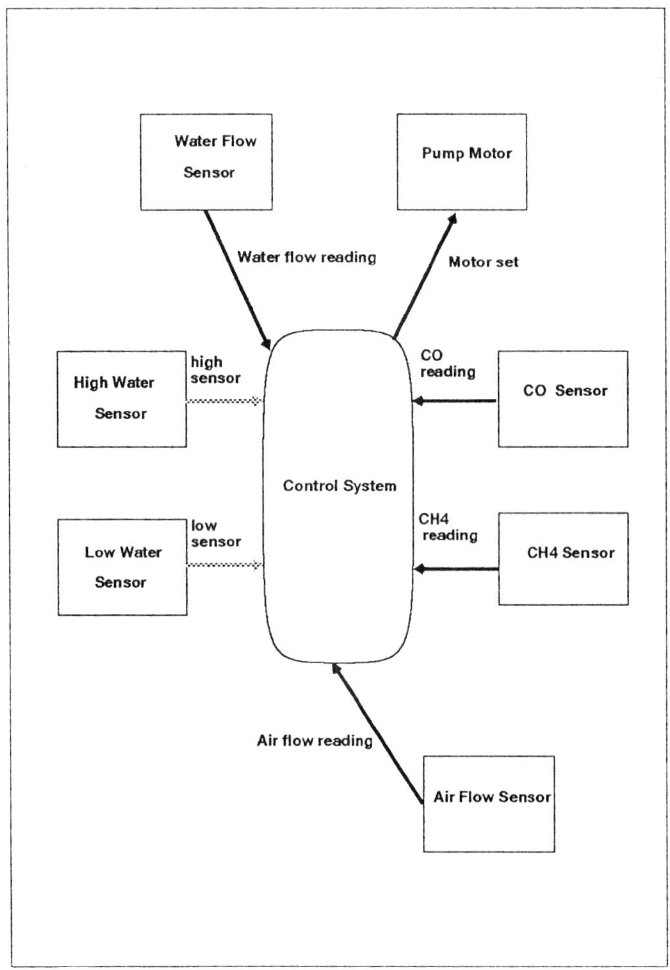

Figure 9.2: Graph Showing External Devices

During our discussions of the case study we shall assume that all devices are controlled in a similar fashion through standard device registers. The interface should ideally be defined by a *passive* object but for simplicity we shall not give its definition. Appendix A presents the full definition of the "device register" object.

9.1.1. Functional Requirements

The functional specification of the system may be divided into four components: the pump operation, the environment monitoring, the operator interaction, and system monitoring.

Pump operation

The required behaviour of the pump is that it monitors the water levels in the sump. When the water reaches a high level (or when requested by the operator) the pump is turned on and the sump is drained until the water reaches the low level. At this point (or when requested by the operator) the pump is turned off. A flow of water in the pipes can be detected if required.

The pump should only be allowed to operate if the methane level in the mine is below a critical level.

Environment monitoring

The environment must be monitored to detect the level of methane in the air; there is a level beyond which it is not safe to cut coal or operate the pump. The monitoring also measures the level of carbon monoxide in the mine and detects whether there is an adequate flow of air. Alarms must be signalled if gas levels become critical.

Operator Interaction

The system is controlled from the surface via an operator's console. The operator is informed of all critical events.

System monitoring

All the system events are to be stored in an archival database, and may be retrieved and displayed upon request.

9.1.2. Non-functional Requirements

The non-functional requirements can be divided into three components: timing, dependability, and security. This case study is mainly concerned with the timing requirements and consequently we shall not address dependability and security (see Burns and Lister[25] for a full consideration of dependability and security aspects).

There are several requirements which relate to the timeliness of system actions. The following list is adapted from Burns and Lister:[25]

(i) Monitoring periods

The maximum periods for reading the environment sensors (see above) may be dictated by legislation. For the purpose of this example we assume these periods are the same for all sensors, namely 60 seconds. In the case of methane there may be a more stringent requirement based on the proximity of the pump and the need to ensure that it never operates when the methane level is critically high. This is discussed in (ii) below.

We assume that the water level detectors are event driven, and that the system should respond within 20 seconds.

(ii) Shut-down deadline

To avoid explosions there is a deadline within which the pump must be switched off once the methane level exceeds a critical threshold. This deadline is related to the methane sampling period, to the rate at which methane can accumulate, and to the margin of safety between the level of methane regarded as critical and the level at which it explodes. The relationship can be expressed by the inequality

$$R(P + D) < M$$

where

R is the rate at which methane can accumulate
P is the sampling period
D is the shut-down deadline
M is the safety margin.

Note that the period P and the deadline D can be traded off against each other, and both can be traded off against the safety margin M. The longer the period or the deadline the more conservative must be the safety margin; the shorter the period or deadline the closer to its safety limits the mine can operate. The designer may therefore vary any of D, P or M in satisfying the deadline and periodicity requirements.

In this example we shall assume that the presence of methane pockets may cause levels to rise rapidly, and therefore we require a sampling period of 5 seconds and a shut-down deadline of 1 second.

(iii) Operator information deadline

The operator must be informed: within 1 second of detection of critically high methane or carbon monoxide readings, within 2 seconds of a critically low air-flow reading, and within 3 seconds of a failure in the operation of the pump.

In summary the sensors have the following defined periods or minimum inter-arrival rates (in seconds) and deadlines.

	periodic/sporadic	arrival times	deadline
CH4_Sensor	P	5.0	1
CO_Sensor	P	60.0	1
Water_Flow	P	60.0	3
Air_Flow	P	60.0	2
Water level detectors	S	100.0	20

Figure 9.3: Attributes of Periodic and Sporadic Processes

9.2. The Logical Architecture Design

We now develop a logical architecture for the pump control system. In the logical architecture we address those requirements which are independent of the physical constraints (eg. processor speeds) imposed by the execution environment. The functional requirements identified in the previous section fall into this category. Consideration of the other system requirements is deferred until design of the physical architecture, described later.

9.2.1. First Level Decomposition

The first step in developing the logical architecture is the identification of appropriate classes of object from which the system can be built. The functional requirements of the system suggest four distinct subsystems:

(i) pump controller subsystem, responsible for operating the pump;

(ii) environment monitor subsystem, responsible for monitoring the environment;

(iii) operator console subsystem, the interface to human operators;

(iv) data logger subsystem, responsible for logging operational and environmental data.

Figure 9.4 illustrates this decomposition. The pump controller has four operations: The operations "not safe" and "is safe" are called by the environment monitor to indicate to the pump controller whether it is safe to operate the pump (due to the level of methane in the environment). The "request status" and "set pump" operations are called by the operator console. As an additional reliability feature the pump controller will always check that the methane level is low before starting the pump (by calling "check safe" in the environment monitor). If pump controller finds that the pump cannot be started (or that the water does not appear to be flowing when the pump is notionally on) then it raises an operator alarm.

The environment monitor has the single operation "check safe" which is called by the pump controller.

The operator console has the alarm operation, which as well as being called by the pump controller, is also called by the environmental monitor if any of its readings are too high. As well as receiving the alarm calls, the operator console can request the status of the pump and attempt to override the high and low water sensors by directly operating the pump. However in the latter case the methane check is still made, with an exception being used to inform the operator that the pump cannot be turned on.

The data logger has six operations which are merely data logging actions which are called by the pump controller and the environment monitor.

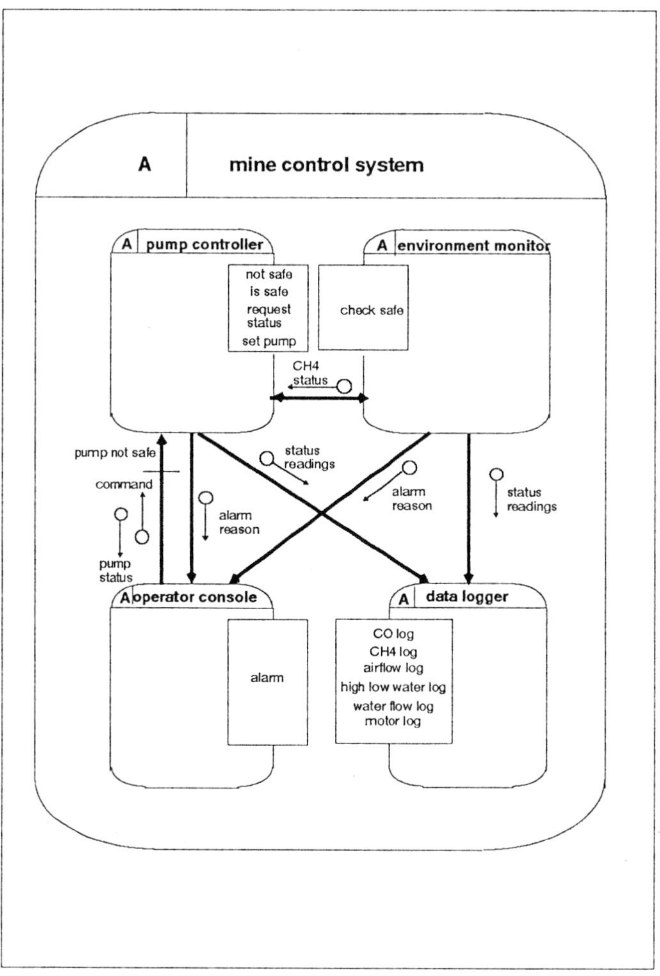

Figure 9.4: First Level Hierarchical Decomposition of Control System

9.2.2. Pump Controller

The decomposition appropriate to the pump controller is shown in Figure 9.5. The pump controller is decomposed into three objects. The first object controls the pump motor. As this object simply responds to commands, requires mutual exclusion for its operations, and does not spontaneous call other objects, it is a *protected* object. All of the pump controller's operations are implemented by the motor object. As the system is real-time, we have attempted to minimise the blocking time of all operations. In this case none of the operations can be blocked. The motor object will make calls to all its uncle objects.

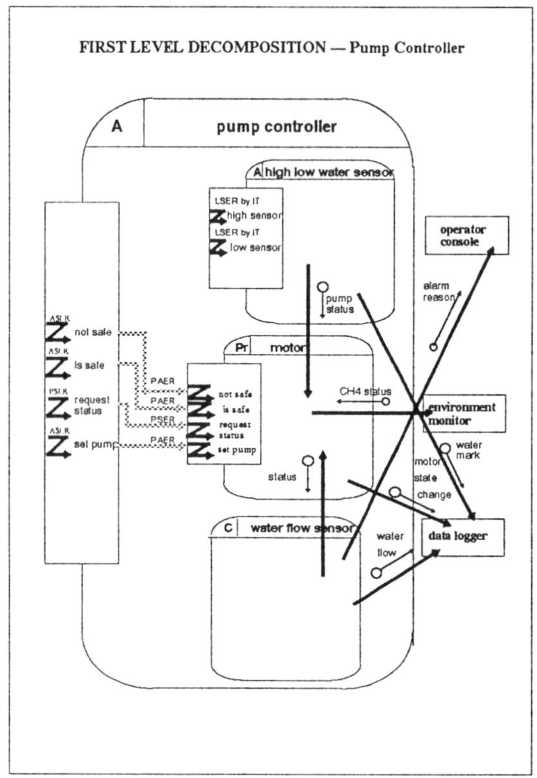

Figure 9.5: Hierarchical Decomposition of the Pump Object

The other two objects control the water sensors. The flow sensor object is a *cyclic* object which continually monitor the flow of water from the mine. The high low water sensor is an *active* object which handles the interrupts from the high low water sensors. It decomposes into a *protected* and a *sporadic* object, as shown in Figure 9.6.

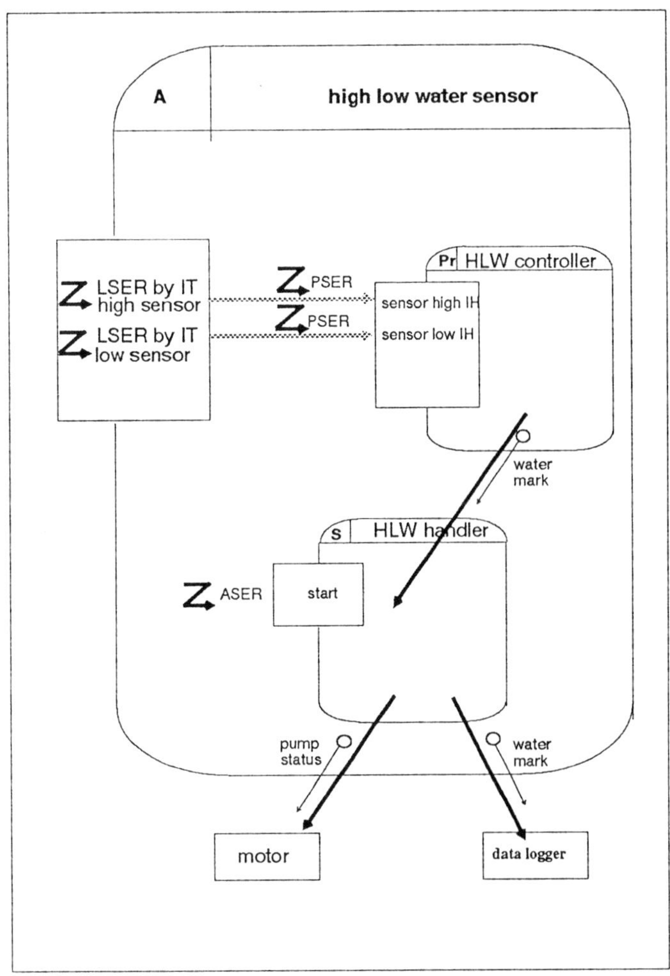

Figure 9.6: Decomposition of the High Low Water Sensor

9.2.3. The Environment Monitor

The Environment Monitor decomposes into four terminal objects, as shown in Figure 9.7. Three of the objects are *cyclic* objects which monitor the: CH4 level, CO level and the air flow in the mine environment. Only the CH4 level is requested by other objects in the system, consequently a *protected* object is used to control access to the current value.

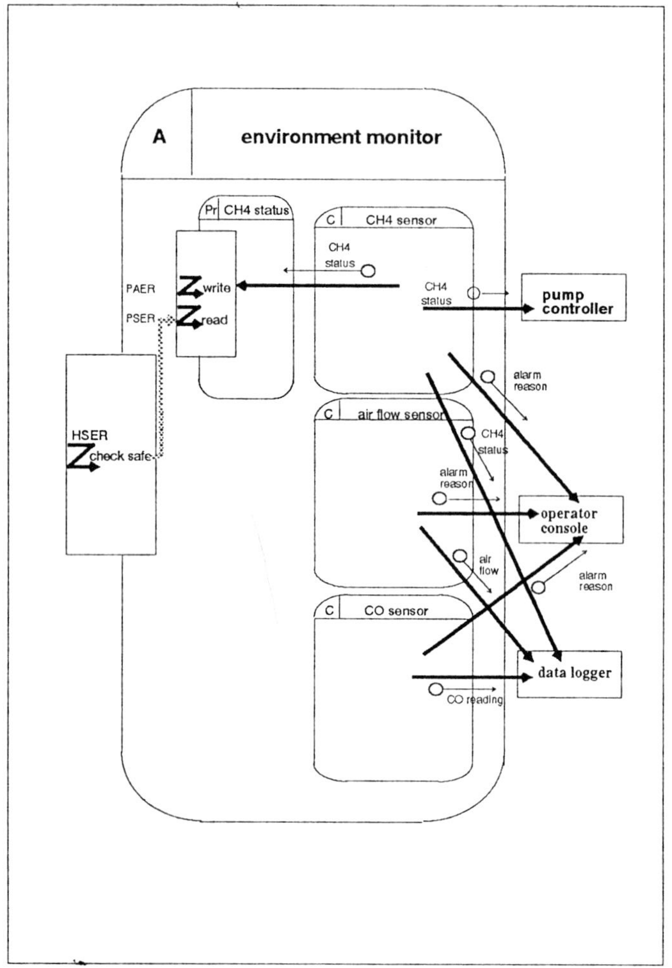

Figure 9.7: Hierarchical Decomposition of the Environment Monitor

9.2.4. The Data Logger

We are not concerned with the details of the data logger in this case study. However, we do require that it presents an asynchronous interface. This is illustrated in Figure 9.8.

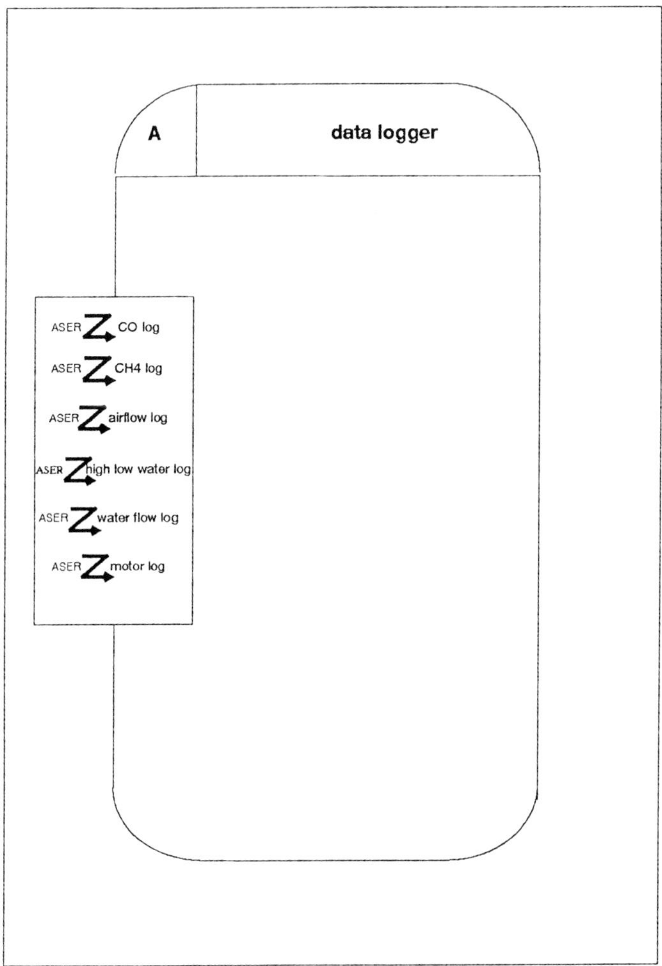

Figure 9.8: The Data Logger

9.2.5. The Operator Console

We are not concerned with the details of the operator console in this case study. However, we do require that it presents an asynchronous interface. This is illustrated in Figure 9.9.

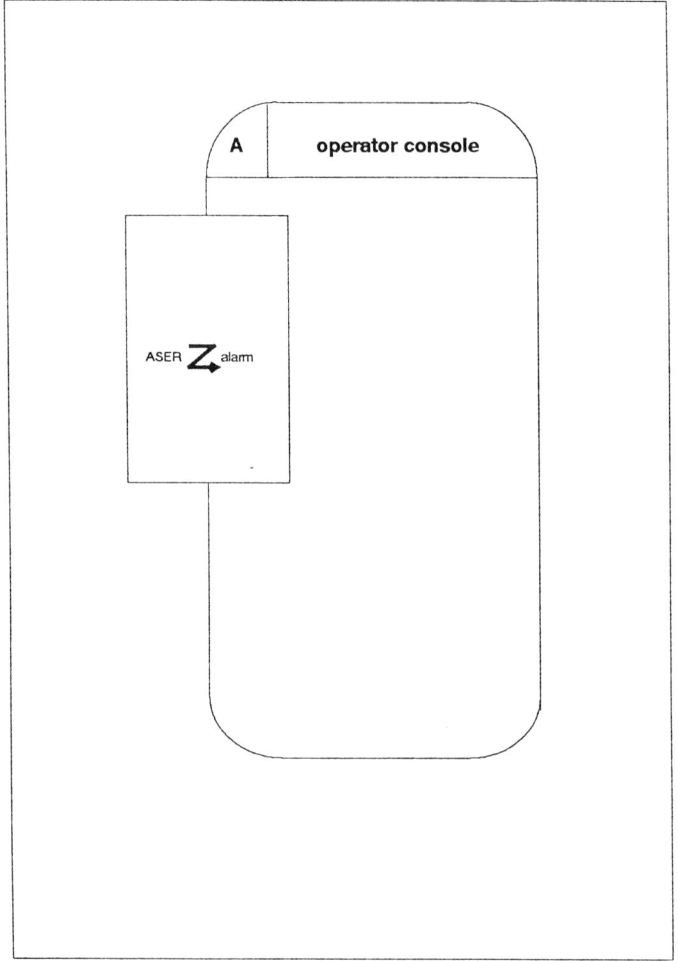

Figure 9.9: The Operator Console

9.3. The Physical Architecture Design

HRT-HOOD supports the design of a physical architecture by:

1) allowing timing attributes to be associated with objects,
2) providing a framework from within which a schedulability analysis of the terminal objects can be undertaken, and
3) providing the abstractions with which the designer can express the handling of timing errors.

The non-functional timing requirements identified in Section 9.1.2 are transformed into annotations on methods and threads as follows.

(i) Periodicity

The threads in the environment monitor which read the carbon monoxide, water flow and air-flow sensors have periods of 60 seconds. The methane monitor thread has a period of 5 seconds in order that it can turn off a running pump within the required deadline.

(ii) Deadlines

The threads which read the environment sensors have a range of deadline (from 1 to 3 seconds) for reporting critical readings to the operator. There is also a deadline of 1 second for executing the pump controller method which switches the pump off when a critical methane level is detected. Figure 9.10 summarises the timing requirements.

	periodic/sporadic	arrival times	deadline
CH4_Sensor	P	5.0	1
CO_Sensor	P	60.0	1
Water_Flow	P	60.0	3
Air_Flow	P	60.0	2
Water level detectors	S	100.0	20

Figure 9.10: Summary of the Timing Characteristics of Threads

9.3.1. Execution Time Analysis

For the purpose of the case study we will assume the execution times given in Figure 9.11 (in practice these would be derived from considering the pseudo code which had been developed with the ODS for each object; we shall miss out this stage and present the full ODSs in the next section). For simplicity we shall ignore the handling of timing errors.

We assume that the high and low water sensor interrupts occur at a priority higher than any software task, and that the low level interrupt handling time is 0.01 seconds. All other system overheads, such as context switch times, have been incorporated into the thread and protected objects worst case execution time.

operation	budget time	error time	handling	wcet
motor				
not safe	0.05		0	0.05
is safe	0.05		0	0.05
request status	0.05		0	0.05
set pump	0.10		0	0.10
Flow sensor				
thread	0.15		0	0.15
HLW_controller				
sensor_high_IH	0.01		0	0.01
sensor_low_IH	0.01		0	0.01
HLW_handler				
start	0.01		0	0.01
thread	0.10		0	0.10
ch4_status				
read	0.01		0	0.01
write	0.01		0	0.01
ch4_sensor				
thread	0.25		0	0.25
air_flow_sensor				
thread	0.15		0	0.15
co_sensor				
thread	0.15		0	0.15

Figure 9.11: Summary of the Timing Characteristics of Operations

During the analysis of the pseudo code the problem of delays inside device drivers must be accounted for. Analogue to digital converters work in many different ways and have a range of operational characteristics. We assume a simple converter which will perform the conversion on request (by successive approximation). This typically might take a few microseconds. A driver task could loop around waiting for the done bit on the device to be set. However, to protect the software from a faulty device (i.e. one that never sets the done bit) the thread is required to delay for 0.1 seconds and then check the flag. If we allow other threads to run during this delay, then extra context switches will be incurred; furthermore there is the possibility that low priority threads might execute and obtain resources (protected objects). This would require the schedulability analysis to take into account this extra blocking time.

For the above reason HRT-HOOD does not allow the application programmer to use arbitrary delays. This forces the programmer to busy wait on the clock. That is, for a delay of 0.1 seconds the programmer must write:

```
NOW := SECONDS(CLOCK);
loop
   exit when SECONDS(CLOCK) >= NOW + 0.1;
end loop;
```

The delay is therefore accounted for by adding to the worst case execution time of the thread. If a larger delay is required then the object must be split into two related ones (with an offset between their executions).

As a result of running the analyser tool we determine that the system is schedulable on a single processor with the following priority settings (13 is highest priority) assuming that priority level 8 represents the hardware priorities of the high low water mark interrupts

HLW_handler_obcs	10
HLW_controller	9
interrupt	8
ch4_status	7
motor	6
ch4_sensor	5
co_sensor	4
air_flow_sensor	3
water_flow_sensor	2
HLW_handler	1

Figure 9.12: Summary of the Priorities

Assuming all threads are released at time 0, and we have preemptive priority-based run-time scheduling, the following diagram (produced by a scheduling simulator called STRESS[8]) illustrates the execution of the threads. The first 100 ticks are where there is most contention, and it can be seen that all threads comfortably meet their deadlines.

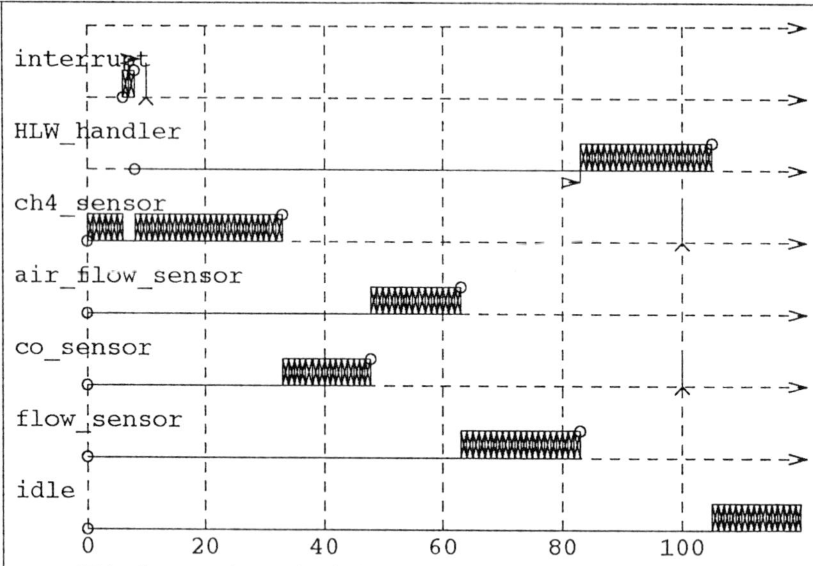

This diagram shows the following aspects of a thread set simulation:

- Thread Release - circle on the time line of a thread (see air_flow_sensor at time 0).
- Releasing a thread blocked on a *protected* object - left corner of a square with a triangle (see interrupt at t = 7).
- Start of execution of a thread previously blocked on a *protected* object - right corner of a square with a triangle (see HLW_handler at t = 82).
- Thread Execute - a hatched box on the time line of a thread (see CH4_sensor at time 0 to 6).
- Preempted Thread - when a higher priority thread runs and a lower priority thread has remaining computation time, a dotted line extend along the time line of the thread (see CH4sensor at time 6 to 8).
- Thread deadline - a vertical line with a triangle at its base (see ch4_sensor at t = 100).
- Missed Deadline - raised circle above the time line of a thread (not shown in the diagram)

Figure 9.13: Simulated Execution of the Mine Control System

9.4. The Object Description Skeleton

Having presented the logical and physical design of the system we undertake detail system design. This is achieved by presenting the object description skeletons. In what follows we assume the existence of a passive object which defines the interface between the hardware devices. Appendix A fully defines this device_register object.

9.4.1. Pump Controller

```
OBJECT Pump_Controller IS ACTIVE
  DESCRIPTION
    This is an object which controls the pump station.
    It encapsulates the pump motor, and various water sensor
    devices
  IMPLEMENTATION_CONSTRAINT
    This is a non-terminal object.
  PROVIDED_INTERFACE
    TYPES

      Pump_Status is (On, Off, Disabled);
      Motor_State_Changes is (Motor_Started, Motor_Stopped,
                              Motor_Safe, Motor_Unsafe);
      Water_Mark is (High, Low);
      Water_Flow is (Yes, No);

    CONSTANTS
      NONE
    OPERATION_SETS
      NONE
    OPERATIONS

      procedure Not_Safe;
      procedure Is_Safe;
      function Request_Status return Pump_Status;
      procedure Set_Pump(To : Pump_Status);

    EXCEPTIONS
      Pump_Not_Safe RAISED_BY Set_Pump;
  REQUIRED_INTERFACE
    OBJECTS

      Environment_Monitor

        CONSTANTS
          NONE
        TYPES

          Methane_Status;

        OPERATIONS

          Check_Safe return Methane_Status;

        EXCEPTIONS
          NONE
```

```
Data_Logger
```

CONSTANTS
  NONE
TYPES
  NONE
OPERATIONS

```
Water_Flow_Log(Reading : Water_Flow);
High_Low_Water_Log(Mark : Water_Mark);
Motor_Log(State : Motor_State_Changes);
```

EXCEPTIONS
  NONE

```
Operator_Console
```

CONSTANTS
  NONE
TYPES

```
Alarm_Reason;
```

OPERATIONS

```
Alarm(Reason: Alarm_Reason)
```

EXCEPTIONS
  NONE

DATAFLOW

```
Pump_Status => Environment_Monitor;
Methane_Status <= Environment_Monitor;
Pump_Status => Data_Logger;
Water_Flow  => Data_Logger;
Water_Mark => Data_Logger;
Motor_State_Changes => Data_Logger;
Alarm_Reason => Operator_Console;
```

EXCEPTION_FLOW

```
Pump_Not_Safe => Operator_Console
```

ACTIVE_OBJECT_CONTROL_STRUCTURE
DESCRIPTION
All operations are HSER with no functional activation constraints.
CONSTRAINED_OPERATIONS
`Not_Safe` CONSTRAINED_BY ASER
`Is_Safe` CONSTRAINED_BY ASER
`Set_Pump` CONSTRAINED_BY ASER
`Request_Status` CONSTRAINED_BY HSER

```
INTERNALS
  OBJECTS

    High_Low_Water_Sensor
    Motor
    Flow_Sensor

  TYPES
    Pump_Status IMPLEMENTED_BY Motor.Pump_Status
    Motor_State_Changes IMPLEMENTED_BY Motor.Pump_Status
    Water_Flow IMPLEMENTED_BY Flow_Sensor.Water_Flow
    Water_Mark IMPLEMENTED_BY
          High_Low_Water_Sensor.Water_Flow
  OPERATIONS
    Not_Safe IMPLEMENTED_BY Motor.Not_Safe
    Is_Safe IMPLEMENTED_BY Motor.Is_Safe
    Request_Status IMPLEMENTED_BY Motor.Request_Status
    Set_Pump IMPLEMENTED_BY Motor.Set_Pump
  EXCEPTIONS
    Pump_Not_Safe IMPLEMENTED_BY Motor.Pump_Not_Safe
  ACTIVE_OBJECT_CONTROL_STRUCTURE
    IMPLEMENTED_BY

      Motor

END pump_controller
```

9.4.2. Motor

```
OBJECT Motor IS PROTECTED
  DESCRIPTION
```

The motor responds to its operations by changing the state of the pump. If the request is for the motor to be turned on (and it is not currently working) then a check is made of the environmental monitor's reading of the methane level. Only if this value is acceptable is the physical motor started.

The environmental monitor will call the motor object when the air becomes unsafe. If the motor is currently running it is immediately turned off. At a later time if "is_safe" is called then the pump will be returned to its original state. If it was previously running then it will be turned on again; if not then its state will change to "off". As with the other objects error conditions are passed on as alarms and data logging activities are programmed.

```
  REAL_TIME_ATTRIBUTES
    CEILING_PRIORITY
      7
    OPERATION Not_Safe IS
      WCET
        0.05
```

```
    BUDGET
      0.05
  END OPERATION Not_Safe

  OPERATION Is_Safe IS
    WCET
      0.05
    BUDGET
      0.05
  END OPERATION Is_Safe

  OPERATION Request_Status IS
    WCET
      0.05
    BUDGET
      0.05
  END OPERATION Request_Status

  OPERATION Set_Pump IS
    WCET
      0.10
    BUDGET
      0.10
  END OPERATION Set_Pump
PROVIDED_INTERFACE
  TYPES

    Pump_Status is (On, Off, Disabled);
    Motor_State_Changes is (Motor_Started, Motor_Stopped,
                            Motor_Safe, Motor_Unsafe);

  CONSTANTS
    NONE
  OPERATION_SETS
    NONE
  OPERATIONS

    procedure Not_Safe
    procedure Is_Safe;
    function Request_Status return  Pump_Status;
    procedure Set_Pump(To: Pump_Status);

  EXCEPTIONS

    Pump_Not_Safe Raised_By Set_Pump;
```

•

REQUIRED_INTERFACE
OBJECTS

```
Environment_Monitor
```

CONSTANTS
NONE
TYPES

```
Methane_Status
```

OPERATIONS

```
Check_Safe return Methane_Status;
```

EXCEPTIONS
NONE

```
Data_Logger
```

CONSTANTS
NONE
TYPES
NONE
OPERATIONS

```
Data_Logger.Motor_Log(State : Motor_State_Changes);
```

EXCEPTIONS
NONE

DATAFLOW

```
Methane_Status <= Environment_Monitor;
Motor_State_Change => Data_Logger;
Pump_Status => Flow_Sensor;
Pump_Status <= High_Low_Water_Sensor;
```

EXCEPTION_FLOW

```
Pump_Not_Safe => Operator_Console
```

PROTECTED_OPERATION_CONTROL_STRUCTURE
DESCRIPTION
All the operation are constrained only by the need for mutual exclusion. There are no functional activation constraints.
CONSTRAINED_OPERATIONS
Not_Safe CONSTRAINED_BY PAER
Is_Safe CONSTRAINED_BY PAER
Set_Pump CONSTRAINED_BY PAER
Request_Status CONSTRAINED_BY PSER

INTERNALS
  OBJECTS
    NONE
  TYPES
    NONE
  DATA

```
Motor_State : Pump_Status := Off;
Return_Condition: Pump_Status := Off;
-- define control and status register
-- for the motor
Pcsr : Device_Register.Csr :=
        (Error_Bit => Clear, Operation => Set,
         Done => False, Interrupt => I_Enabled,
         Device => D_Enabled);
for Pcsr'Address use 16#Aa14#;
```

  OPERATIONS
    NONE

OPERATION_CONTROL_STRUCTURE
  OPERATION Not_Safe IS PROTECTED
    DESCRIPTION
      The procedure is called when the methane level is such that it is no longer safe to operate the pump.
    USED_OPERATIONS

```
Data_Logger.Motorlog
```

    PROPAGATED_EXCEPTIONS
      NONE
    HANDLED_EXCEPTIONS
      NONE
    CODE

```
if Motor_Status /= Disabled then
  Return_Condition := Motor_Status;
  Motor_Status := Disabled;
  Pcsr.Operation := Clear; -- turn off motor
  Data_Logger.Motor_Log(Motor_Unsafe);
end if;
```

  END OPERATION Not_Safe

  OPERATION Is_Safe IS PROTECTED
    DESCRIPTION
      This procedure is called when the methane level is such that it is safe to operate the pump.

USED_OPERATIONS

```
Data_Logger.Motor_Log
```

PROPAGATED_EXCEPTIONS
NONE
HANDLED_EXCEPTIONS
NONE
CODE

```
Data_Logger.Motor_Log(Motor_Safe);
if Return_Condition = On then
  Pcsr.Operation := Set; -- start motor
  Data_Logger.Motor_Log(Motor_Started);
end if;
Motor := Return_Condition;
```

END OPERATION `Is_Safe`

OPERATION `Request_Status Return Pump_Status` IS PROTECTED
DESCRIPTION
This function returns the current status of the pump motor.
USED_OPERATIONS

```
Data_Logger.Motor_Log
```

PROPAGATED_EXCEPTIONS
NONE
HANDLED_EXCEPTIONS
NONE
CODE

```
return Motor_Status;
```

END OPERATION `Request_Status`

OPERATION `Set_Pump(To: Pump_Status)` IS PROTECTED
DESCRIPTION
This procedure sets the pump to a given status.
USED_OPERATIONS

```
Data_Logger.Motor_Log;
Environment_Monitor.Check_Safe;
```

PROPAGATED_EXCEPTIONS

```
$Pump_Controller.Pump_Not_Safe$;
```

HANDLED_EXCEPTIONS
NONE

CODE

```
if To = On then
  if Motor_Status = Off or
           Motor_Status = Disabled then
    if $Environment_Monitor.Check_Safe$ =
           Motor_Safe then
      Motor_Status := On;
      Pcsr.Operation := Set; -- turn on motor
      Data_Logger.Motor_Log(Motor_Started);
    elsif Motor_Status = Disabled then
      raise $Pump_Controller.Pump_Not_Safe$;
    end if;
  end if;
else -- to = off
  if Motor_Status = On then
    Pcsr.Operation := Clear; -- turn off motor
    Motor_Status:= Off;
    Data_Logger.Motor_Log(Motor_Stopped);
  end if;
end if;
```

END OPERATION Set_Pump

PROTECTED_OBJECT_CONTROL_STRUCTURE

CODE

All operations only require mutual exclusion, there is no synchronisation code.

END_OBJECT Motor

9.4.3. Flow Sensor

OBJECT Flow_Sensor IS CYCLIC

DESCRIPTION

This object monitors the flow of water from the mine. It is used to check that the pump is working.

REAL_TIME_ATTRIBUTES

PERIOD

60.0

OFFSET

0

DEADLINE

3

PRIORITY

6

IMPORTANCE

Hard

```
  OPERATION Thread IS
    WCET
      0.15
    BUDGET
      0.15
  END_OP Thread

PROVIDED_INTERFACE
  TYPES

    Water_Flow is (Yes, No);

  CONSTANTS
    NONE
  OPERATIONS
    NONE
  OPERATION_SETS
    NONE
  EXCEPTIONS
    NONE

REQUIRED_INTERFACE
  OBJECTS

    Motor

      CONSTANTS
        NONE
      TYPES

        Pump_Status;

      OPERATIONS

        Request_Status return Pump_Status;

      EXCEPTIONS
        NONE

    Operator_Console

      CONSTANTS
        NONE
      TYPES

        Alarm_Reason;

      OPERATIONS

        Alarm(Reason: Alarm_Reason);

      EXCEPTIONS
        NONE
```

```
Data_Logger
```

CONSTANTS
NONE
TYPES

```
Water_Flow;
```

OPERATIONS

```
Water_Flow Log(Reading : Water_Flow);
```

EXCEPTIONS
NONE

DATAFLOW

```
Pump_Status <= Motor;
Alarm_Reason => Operator_Console;
Water_Flow => Data_Logger;
```

EXCEPTION_FLOW
NONE

CYCLIC_OPERATION_CONTROL_STRUCTURE
DESCRIPTION
As there are no operations to this object, there is no OBCS.
CONSTRAINED_OPERATIONS
NONE
HANDLED_EXCEPTIONS
NONE

INTERNALS
OBJECTS
NONE
TYPES
NONE
DATA

```
Flow : Water_Flow := Off;
Current_Pump_Status, Last_Pump_Status : Pump_Status;
-- define control and status register
-- for the flow switch
Wfcsr : Device_Register.Csr(Device => D_Enabled);
for Wfcsr'Address use 16#Aa14#;
```

OPERATIONS

```
Initialise
```

OPERATION_CONTROL_STRUCTURE
OPERATION `Initialise` IS
DESCRIPTION
Initialisation code
USED_OPERATIONS
NONE
PROPAGATED_EXCEPTIONS
NONE
HANDLED_EXCEPTIONS
NONE
CODE

```
-- enable device
Wfcsr.Device := D_Enabled;
```

END OPERATION `Initialise`

OPERATION `Thread` IS
DESCRIPTION
This task is a standard example of a periodic process. During each period it checks to see if the motor is running and the water flowing. As the water sensor is some distance from the pump, it is possible for the pump to have just started running but that no water has reached the sensor. To cater for this event a delay is used before a second reading of the water flow is made. "Alarm" is called only if there is no flow when the second reading is taken.
USED_OPERATIONS

```
Operator_Console.Alarm;
Data_Logger.Water_Flow_Log;
Motor.Request_Status;
```

PROPAGATED_EXCEPTIONS
NONE
HANDLED_EXCEPTIONS
NONE
CODE

```
Current_Pump_Status := Motor.Request_Status;
```

```
    -- read device register
    if (Wfcsr.Operation = Set) then
        Flow :=  Yes;
    else
        Flow := No;
    end if;
    if Current_Pump_Status = On and
       Last_Pump_Status = On and
                     Flow = No then
       Operator_Console.Alarm(Pump_Dead);
    end if;
    Last_Pump_Status := Current_Pump_Status;

    Data_Logger.Water_Flow_Log(Flow);
```

END OPERATION Thread

CYCLIC_OPERATION_CONTROL_STRUCTURE
CODE
All operations only require mutual exclusion, there is no synchronisation code.
END_OBJECT Flow_Sensor

9.4.4. High Low Water Sensor

OBJECT High_Low_Water_Sensor IS ACTIVE

DESCRIPTION
This is object responds to the high and low water interrupt.

IMPLEMENTATION_CONSTRAINT
Operations are called by hardware interrupts.
PROVIDED_INTERFACE
TYPES

```
Water_Mark is (High, Low)
```

CONSTANTS
NONE
OPERATION_SETS
NONE
OPERATIONS

```
procedure High_Sensor
procedure Low_Sensor
```

EXCEPTIONS
NONE

```
REQUIRED_INTERFACE
  OBJECTS

    Motor

      CONSTANTS
        NONE
      TYPES

        Pump_Status;

      OPERATIONS

        Set_Pump(To : Pump_Status);

      EXCEPTIONS
        NONE

    Data_Logger

      TYPES
        NONE
      OPERATIONS

        High_Low_Water_Log(Mark : Water_Mark);

      EXCEPTIONS
        NONE

DATAFLOW

  Pump_Status => Motor;
  Water_Mark => Data_Logger;

EXCEPTION_FLOW
  NONE

ACTIVE_OBJECT_CONTROL_STRUCTURE
  DESCRIPTION
    The two operations are mutually exclusive.
  CONSTRAINED_OPERATIONS
    Sensor_High CONSTRAINED_BY LSER_BY_INTERRUPT
    Sensor_Low CONSTRAINED_BY LSER_BY_INTERRUPT

INTERNALS
  OBJECTS

    Hlw_Controller, Hlw_Handler

  TYPES
    Water_Mark IMPLEMENTED_BY Hlw_Handler.Water_Mark
```

```
  OPERATIONS
    Sensor_High IMPLEMENTED_BY Hlw_Controller.Sensor_High.Ih
    Sensor_Low IMPLEMENTED_BY Hlw_Controller.Sensor_Low

ACTIVE_OBJECT_CONTROL_STRUCTURE
  IMPLEMENTED_BY

    Hlw_Controller

END_OBJECT High_Low_Water_Sensor
```

9.4.5. Controller

```
OBJECT Hlw_Controller IS PROTECTED

  DESCRIPTION
    This is object handles the actual interrupts and interfaces to the handling
    sporadic.

  REAL_TIME_ATTRIBUTES
    CEILING_PRIORITY
      13
    OPERATION Sensor_High_Ih IS
      WCET
        0.01
      BUDGET
        0.01
    END OPERATION Sensor_High_Ih

    OPERATION Sensor_Low_Ih IS
      WCET
        0.01
      BUDGET
        0.01
    END OPERATION Sensor_Low_Ih

  PROVIDED_INTERFACE
    TYPES
      NONE
    CONSTANTS
      NONE
    OPERATIONS

      procedure High_Sensor_Ih
      procedure Low_Sensor_Ih

    EXCEPTIONS
      NONE
```

REQUIRED_INTERFACE
OBJECTS

```
Hlw_Handler
```

CONSTANTS
NONE
TYPES

```
Water_Mark
```

OPERATIONS

```
Start(Int : Water_Mark);
```

EXCEPTIONS
NONE

DATAFLOW

```
Water_Mark => Hlw_Handler;
```

EXCEPTION_FLOW
NONE

PROTECTED_OPERATION_CONTROL_STRUCTURE
DESCRIPTION
There are two interrupts which must be handled, consequently the sporadic thread can be invoked from two sources. The objects obcs controls this synchronisation.
CONSTRAINED_OPERATIONS
`Sensor_High_Ih` CONSTRAINED_BY PSER
`Sensor_Low_Ih` CONSTRAINED_BY PSER

INTERNALS
OBJECTS
NONE
TYPES
NONE
OPERATIONS
NONE

OPERATION_CONTROL_STRUCTURE
OPERATION `Sensor_High_Ih` IS PROTECTED
DESCRIPTION
This is the handling routine for the high water sensor.
USED_OPERATIONS

```
Hlw_Handler.Start
```

PROPAGATED_EXCEPTIONS

NONE
HANDLED_EXCEPTIONS
NONE
CODE

```
Hlw_Handler.Start(High);
```

END OPERATION `Sensor_High_Ih`

OPERATION `Sensor_Low_Ih` IS PROTECTED
DESCRIPTION
This is the handling routine for the low water sensor.
USED_OPERATIONS

```
Hlw_Handler.Start
```

PROPAGATED_EXCEPTIONS
NONE
HANDLED_EXCEPTIONS
NONE
CODE

```
Hlw_Handler.Start(Low);
```

END OPERATION `Sensor_Low_Ih`
PROTECTED_OBJECT_CONTROL_STRUCTURE
CODE

```
protected Obcs is
  entry Sporadic_Start (Int : out Water_Mark);
  procedure Sensor_High_Ih;
  pragma Attach_Handler(Sensor_High_Ih, Waterh_Interrupt);
          -- assigns interrupt handler
  procedure Sensor_Low_Ih;
  pragma Attach_Handler(Sensor_Low_Ih, Waterl_Interrupt);
          -- assigns interrupt handler

private
end Obcs;
```

```
protected body Obcs is

  procedure Sensor_High_Ih is
  begin
      -- mutual exclusion only
  end;

  procedure Sensor_Low_Ih is
  begin
     -- mutual exclusion only
  end;

end Obcs;
```

END_OBJECT `Hlw_Controller`

9.4.6. Handler

OBJECT `Hlw_Handler` IS SPORADIC

DESCRIPTION
This is a sporadic object which responds to the high and low water interrupts. Note that it is assumed that these interrupts are mutually exclusive, and therefore it would be possible to estimate a minimum inter-arrival time.

REAL_TIME_ATTRIBUTES
MINIMUM_ARRIVAL_TIME
100.0
OFFSET
0
DEADLINE
20
PRIORITY
4
IMPORTANCE
Hard

OPERATION `Start` IS
WCET
0.01
BUDGET
0.01
END OPERATION `Start`

OPERATION `Thread` IS
WCET
0.10

```
      BUDGET
        0.10
    END OPERATION Thread

  PROVIDED_INTERFACE
    TYPES

      Water_Mark is (High, Low);

    CONSTANTS
      NONE
    OPERATIONS

      procedure Start(Int : Water_Mark);

    EXCEPTIONS
      NONE

  REQUIRED_INTERFACE
    OBJECTS

      Motor

        CONSTANTS
          NONE
        TYPES

          Pump_Status

        OPERATIONS

          Set_Pump(To : Pump_Status);

        EXCEPTIONS
          NONE

      Data_Logger

        TYPES
          NONE
        OPERATIONS

          High_Low_Water_Log(Mark : Water_Mark);

        EXCEPTIONS
          NONE

DATAFLOW

  Water_Mark <= Hlw_Controller;
  Pump_Status => Motor;
  Water_Mark => Data_Logger;
```

EXCEPTION_FLOW

```
Pump_Not_Safe <= Motor
```

SPORADIC_OBJECT_CONTROL_STRUCTURE
  DESCRIPTION
    All communication with this sporadic is asynchronous.
  CONSTRAINED_OPERATIONS
    Start CONSTRAINED_BY ASER;

INTERNALS
  OBJECTS
    NONE
  TYPES

```
-- define control and status registers
-- for the high and low water switches
Hwcsr : Device_Register.Csr;
for Hwcsr'Address use 16#Aa10#;
Lwcsr : Device_Register.Csr;
for Lwcsr'Address use 16#Aa12#;

Int : Water_Mark;
```

  OPERATIONS
    initialise

OPERATION_CONTROL_STRUCTURE
  OPERATION Initialise IS
    DESCRIPTION
      Initialisation operation
    USED_OPERATIONS
      NONE
    PROPAGATED_EXCEPTIONS
      NONE
    HANDLED_EXCEPTIONS
      NONE
    CODE

```
Hwcsr.Device :=D_Enabled;
Hwcsr.Interrupt := I_Enabled;
Lwcsr.Device :=D_Enabled;
Lwcsr.Interrupt := I_Enabled;
```

  END OPERATION Initialise

  OPERATION Thread IS
    DESCRIPTION

The thread simply responds to its invocation request by informing the pump controller when the water is high or low.

USED_OPERATIONS

```
Data_Logger.High_Low_Water_Mark;
Motor.Set_Pump;
```

PROPAGATED_EXCEPTIONS

NONE

HANDLED_EXCEPTIONS

NONE

CODE

```
if Int = High then
  Motor.Set_Pump(On);
  Data_Logger.High_Low_Water_Log(High);
else
  Motor.Set_Pump(Off);
  Data_Logger.High_Low_Water_Log(Low);
end if;
```

END OPERATION Thread

SPORADIC_OBJECT_CONTROL_STRUCTURE

CODE

```
protected Obcs is
  -- for the START operation
  procedure Start(Int : Water_Mark);
  entry Wait_Start(Int : out Water_Mark;
        Start_Time : out Monotonic.Time;
        Overrun : out Boolean);
private
  Start_Open : Boolean := False;
  Soverrun : Boolean := False;
  Freq_Overrun : Boolean := False;
  W : Water_Mark;
  Start_Time : Time;
end Obcs;
```

```
protected body Obcs is

  procedure Start(Int : Water_Mark) do
    W := Int;
    T := Calendar.Time; -- log time of invocation request
    if Start_Open then
      Soverrun := True;
    else
      Start_Time := Clock;
      Start_Open := True;
    end if;
  end Start;

  entry Wait_Start(Int : out Water_Mark;
                   Start_Time : out Monotonic.Time;
                   Overrun : out Boolean)
                   when Start_Open do
    Int := W;
    Started := Start_Time;
    Overrun := Soverrun;
    Soverrun := False;
    Start_Open := False;
  end Wait_Start;

end Obcs;
```

END_OBJECT `Hlw_Handler`

9.4.7. Environment Monitor

OBJECT `Environment_Monitor` IS ACTIVE

DESCRIPTION

This objects monitors the carbon monoxide and methane levels in the mine.

IMPLEMENTATION_CONSTRAINT

This is a non terminal object.

PROVIDED_INTERFACE

TYPES

```
Ch4_Reading is new Integer range 0 .. 1023;
Co_Reading is new Integer range 0 .. 1023;
Methane_Status is (Motor_Safe, Motor_Unsafe);
Air_Flow_Status is (Air_Flow, No_Air_Flow);
```

CONSTANTS

```
Co_High : constant Co_Reading := 600;
Ch4_High : constant Ch4_Reading := 400;
```

OPERATION_SETS
NONE
OPERATIONS

```
function Check_Safe return Methane_Status;
```

EXCEPTIONS
NONE

REQUIRED_INTERFACE
OBJECTS

```
Motor
```

CONSTANTS
NONE
TYPES
NONE
OPERATIONS

```
Not_Safe;
Is_Safe;
```

EXCEPTIONS
NONE

```
Data_Logger
```

TYPES
NONE
OPERATIONS

```
Co_Log(Reading : Co_Reading);
Ch4_Log(Reading : Ch4_Reading);
Air_Log(Reading : Air_Flow_Status);
```

EXCEPTIONS
NONE

```
Operator_Console
```

TYPES

```
Alarm_Reason
```

OPERATIONS

```
Operator_Console.Alarm_Reason;
```

EXCEPTIONS
NONE

DATAFLOW

```
Methane_Status => Pump_Controller;
Alarm_Reason => Operator_Console;
Co_Reading => Data_Logger;
Ch4_Reading => Data_Logger;
Air_Flow_Status => Data_Logger;
```

EXCEPTION_FLOW
NONE

ACTIVE_OBJECT_CONTROL_STRUCTURE
DESCRIPTION
The check_safe operation is HSER constrained with no functional activation constraints.
CONSTRAINED_OPERATIONS
`Check_Safe` CONSTRAINED_BY HSER

INTERNALS
OBJECTS

```
Co_Sensor, Ch4_Sensor, Air_Flow_Sensor, Ch4_Status
```

CONSTANTS

```
Co_High Implemented_By Co_Sensor
Ch4_High Implemented_By Ch4_Sensor
```

TYPES
`Ch4_Reading` IMPLEMENTED_BY `Ch4_Sensor`
`Co_Reading` IMPLEMENTED_BY `Co_Sensor`
`Methane_Status` IMPLEMENTED_BY `Ch4_Status`
`Air_Flow_Status` is IMPLEMENTED_BY `Air_Flow_Sensor`
OPERATIONS
NONE

OPERATION_CONTROL_STRUCTURE
OPERATION `Check_Safe` IMPLEMENTED_BY `Ch4_Status.Read`

ACTIVE_OBJECT_CONTROL_STRUCTURE
IMPLEMENTED_BY

```
Ch4_Status.Read
```

END_OBJECT `Environment_Monitor`

9.4.8. CH4_status

OBJECT Ch4_Status IS PROTECTED

DESCRIPTION
The objects protects the status of the CH4 sensor from concurrent access.
REAL_TIME_ATTRIBUTES

CEILING_PRIORITY
10

OPERATION Write
WCET
0.01
BUDGET
0.01
END OPERATION Write

OPERATION Read

WCET
0.01
BUDGET
0.01
END OPERATION Read

PROVIDED_INTERFACE
TYPES

```
Methane_Status is (Motor_Safe, Motor_Unsafe);
```

CONSTANTS
NONE
OPERATIONS

```
function Read return Methane_Status;
procedure Write(Current_Status : in Methane_Status);
```

EXCEPTIONS
NONE

REQUIRED_INTERFACE
OBJECTS
NONE

DATAFLOW

```
Methane_Status => Motor;
Methane_Status <= Ch4_Sensor;
```

EXCEPTION_FLOW
NONE

PROTECTED_OPERATION_CONTROL_STRUCTURE
DESCRIPTION
Both the operation are constrained only by the need for mutual exclusion. There are no functional activation constraints.
CONSTRAINED_OPERATIONS
`Read` CONSTRAINED_BY PSER
`Write` CONSTRAINED_BY PAER

INTERNALS
OBJECTS
NONE
TYPES
NONE
DATA

```
Environment_Status : Methane_Status := Motor_Unsafe;
```

OPERATIONS
NONE

OPERATION_CONTROL_STRUCTURE
OPERATION `Read Return Methane_Status` IS PROTECTED
DESCRIPTION
Reads the current value of the methane status
USED_OPERATIONS
NONE
PROPAGATED_EXCEPTIONS
NONE
HANDLED_EXCEPTIONS
NONE
CODE

```
return Environment_Status;
```

END OPERATION `Read`

OPERATION Write(Current_Status : Methane_Status)
IS PROTECTED
DESCRIPTION
Updates the current value of the methane status
USED_OPERATIONS
NONE
PROPAGATED_EXCEPTIONS
NONE
HANDLED_EXCEPTIONS
NONE
CODE

```
Environment_Status := Current_Status;
```

END OPERATION Write

PROTECTED_OBJECT_CONTROL_STRUCTURE
CODE
All operations only require mutual exclusion, there is no synchronisation code.
END_OBJECT Ch4_Status

9.4.9. CH4_sensor

OBJECT Ch4_Sensor IS CYCLIC

DESCRIPTION
This periodic process monitors the methane level in the mine.

REAL_TIME_ATTRIBUTES
PERIOD
5
OFFSET
0
DEADLINE
1
PRIORITY
11
IMPORTANCE
Hard
OPERATION Thread
WCET
0.25
BUDGET
0.25
END OPERATION Thread

PROVIDED_INTERFACE
  TYPES

```
Ch4_Reading is new Integer range 0 .. 1023;
```

  CONSTANTS

```
Ch4_High : constant Ch4_Reading := 400;
```

  OPERATIONS
    NONE
  EXCEPTIONS
    NONE

REQUIRED_INTERFACE
  OBJECTS

```
Pump_Controller
```

    CONSTANTS
      NONE
    TYPES
      NONE
    OPERATIONS

```
Is_Safe
Not_Safe
```

    EXCEPTIONS
      NONE

```
Data_Logger
```

    CONSTANTS
      NONE
    TYPES
      NONE
    OPERATIONS

```
Ch4_Log(Reading : Ch4_Reading);
```

    EXCEPTIONS
      NONE

```
Operator_Console
```

    CONSTANTS
      NONE
    TYPES

```
Alarm_Reason;
```

    OPERATIONS

```
Alarm(Reason: Alarm_Reason)
```

```
Ch4_Status
```

CONSTANTS
NONE
TYPES

```
Methane_Status;
```

OPERATIONS

```
Write(Current_Status : Methane_Status)
```

EXCEPTIONS
NONE

DATAFLOW

```
Methane_Status => Ch4_Status
Alarm_Status => Operator_Console
Ch4_Reading => Data_Logger
```

EXCEPTION_FLOW
NONE

CYCLIC_OPERATION_CONTROL_STRUCTURE
DESCRIPTION
As there are no asynchronous transfers of control, there is no OBCS.
CONSTRAINED_OPERATIONS
NONE
HANDLED_EXCEPTIONS
NONE

INTERNALS
OBJECTS
NONE
TYPES

```
Ch4_Present : Ch4_Reading;
-- define control and status register
-- for the CH4 ADC
Ch4csr : Device_Register.Csr;
for Ch4csr'Address use 16#Aa18#;
-- define the data register
Ch4dbr : Ch4_Reading;
for Ch4dbr'Address use 16#Aa1a#;
for Ch4dbr'Size use One_Word;
Jitter_Range : constant Ch4_Reading := 40;
Now : Duration;
```

OPERATIONS

```
Initialise
```

OPERATION_CONTROL_STRUCTURE
OPERATION `Initialise`
DESCRIPTION
The initialisation code
USED_OPERATIONS
NONE
PROPAGATED_EXCEPTIONS
NONE
HANDLED_EXCEPTIONS
NONE
CODE

```
-- enable device
Ch4csr.Device := D_Enabled;
```

END OPERATION `Initialise`

OPERATION `Thread` IS
DESCRIPTION
Analogue to digital converters work in many different ways and have a range of operational characteristics. The code presented here is for a simple converter. Having been asked for a reading (by setting CH4CSR.OPERATION) the device will perform the conversion (by successive approximation), place the scaled integer on the data buffer register and then set the done bit on the control/status register (this typically might take a few microseconds). The driver task loops around waiting for the done bit to be set. Other types of converter are interrupt driven or give a continuous reading in the data buffer. To protect the software from a faulty device (i.e. one that never sets the done bit) the task delays for 0.1 seconds and then checks the flag. The device's operational characteristics dictate that the reading should be available well within the time range. If the done bit is not set then an error condition is reported to the operator console.

Once a reading has been obtained it is compared with the HIGH value and, if necessary, the appropriate call in the pump controller is made. Alternatively if the reading is below the HIGH level, and the motor is disabled (unsafe), then a call to restart the motor is undertaken. To prevent the motor being continually switched on and off when the methane level is hovering around the HIGH level a "jitter_range" (hysteresis) is incorporated into the test for motor_safe.

The above algorithm uses a simple tactic to decide if the methane level is acceptable. Other strategies might involve monitoring the change in methane level so that predictions could be made about the likelihood of a HIGH reading in the future.

Having taken a reading (and undertaken whatever communications with the pump controller are appropriate) the task then logs its reading in the data logger.

USED_OPERATIONS

```
Operator_Console.Alarm;
Pump_Controller.Not_Safe;
Ch4_Status.Read;
Ch4_Status.Write;
Data_Logger.Ch4_Log;
```

PROPAGATED_EXCEPTIONS

NONE

HANDLED_EXCEPTIONS

others

DESCRIPTION

One possible exception that would be caught is CONSTRAINT_ERROR. However, as this is a 10 bit device register which is mapped to the least significant end of a 16 bit memory location the error should not occur, and is therefore not caught explicitly. HANDLED_EXCEPTIONS

CODE

```
Operator_Console.Alarm(Unknown_Error);
Pump_Controller.Not_Safe;
Environment_Status := Motor_Unsafe;
-- try and turn motor off
-- before terminating
```

CODE

```
Ch4csr.Operation := Set; -- start conversion
Now := Seconds(Clock);                  -- wait for conversion
loop
   exit when Seconds(Clock) >= Now + 0.1;
end loop;
```

```
if not Ch4csr.Done then
   Operator_Console.Alarm(Ch4_Device_Error);
else
   -- read device register for sensor value
   Ch4_Present := Ch4dbr;
   if Ch4_Present > Ch4_High then
     if Ch4_Status.Read = Motor_Safe then
       $Pump_Controller.Not_Safe$;
       Ch4_Status.Write(Motor_Unsafe);
     end if;
   elsif (Ch4_Present < (Ch4_High - Jitter_Range))
       and  (Ch4_Status.Read = Motor_Unsafe) then
     $Pump_Controller.Is_Safe$;
     Ch4_Status.Write(Motor_Safe);
   end if;
   Data_Logger.Ch4_Log(Ch4_Present);
end if;
```

END OPERATION Thread

CYCLIC_OPERATION_CONTROL_STRUCTURE
  CODE
    NONE
END_OBJECT Ch4_Sensor

9.4.10. Airflow_sensor

OBJECT Airflow_Sensor IS CYCLIC
  DESCRIPTION
    This periodic process monitors the air flow in the mine.
  REAL_TIME_ATTRIBUTES
    PERIOD
      60
    OFFSET
      0
    DEADLINE
      2
    PRIORITY
      8
    IMPORTANCE
      Hard

    OPERATION Thread
      WCET
        0.15
      BUDGET
        0.15

```
    END OPERATION Thread

  PROVIDED_INTERFACE
    TYPES
      air_flow_status is (air_flow, no_air_flow);
    CONSTANTS
      NONE
    OPERATIONS
      NONE
    EXCEPTIONS
      NONE

  REQUIRED_INTERFACE
    OBJECTS

      Data_Logger

        TYPES
          NONE
        OPERATIONS

          Air_Log(Reading : Air_Flow_Status);

        EXCEPTIONS
          NONE

      Operator_Console

        TYPES

          Alarm_Reason

        OPERATIONS

          Operator_Console.Alarm_Reason;

        EXCEPTIONS
          NONE

DATAFLOW

  Alarm_Reason => Operator_Console;
  Air_Flow_Status => Data_Logger;


EXCEPTION_FLOW
  NONE

CYCLIC_OPERATION_CONTROL_STRUCTURE
  DESCRIPTION
    As there are no asynchronous transfers of control, there is no OBCS.
  CONSTRAINED_OPERATIONS
```

NONE
HANDLED_EXCEPTIONS
NONE

INTERNALS
OBJECTS
NONE
TYPES
NONE
DATA

```
Air_Flow_Reading : Boolean;
-- define control and status register
-- for the flow switch
Afcsr : Device_Register.Csr;
for Afcsr'Address use 16#Aa20#;
```

OPERATIONS

```
Initialise
```

OPERATION_CONTROL_STRUCTURE
OPERATION `Initialise` IS
DESCRIPTION
The initialisation code
USED_OPERATIONS
NONE
PROPAGATED_EXCEPTIONS
NONE
HANDLED_EXCEPTIONS
NONE
CODE

```
-- enable device
Afcsr.Device := D_Enabled;
```

END OPERATION `Initialise`

OPERATION `Thread` IS
DESCRIPTION
The air flow sensor is also very simple, it needs a long sample time and only signals a lack of air flow if no current has been detected over this period. It signals the lack of air flow via the set bit in the control/status buffer. The controller task merely checks this value and calls alarm if no flow is observed.
USED_OPERATIONS

```
      Operator_Console.Alarm;
      Data_Logger.Air_Flow_Log
    PROPAGATED_EXCEPTIONS
      NONE
    HANDLED_EXCEPTIONS
      NONE
    CODE
      -- read device register for flow indication
      -- (operation bit set to 1);
      Air_Flow_Reading := Afcsr.Operation = Set;
      if not Air_Flow_Reading then
         Operator_Console.Alarm(No_Air_Flow);
            Data_Logger.Air_Flow_Log(No_Air_Flow);
          else
            Data_Logger.Air_Flow_Log(Air_Flow);
      end if;
  END OPERATION Thread

CYCLIC_OPERATION_CONTROL_STRUCTURE
  CODE
    NONE
END_OBJECT Airflow_Sensor
```

9.4.11. CO_sensor

```
  OBJECT Co_Sensor CYCLIC
    DESCRIPTION
      The objects monitors the carbon monoxide level in the mine.
    REAL_TIME_ATTRIBUTES
      PERIOD
        60
      OFFSET
        0
      DEADLINE
        1
      PRIORITY
        10
      IMPORTANCE
        Hard

      OPERATION Thread IS
        WCET
          0.15
        BUDGET
          0.15
```

```
    END OPERATION Thread

  PROVIDED_INTERFACE
    TYPES

      Co_Reading is new Integer range 0 .. 1023;

    CONSTANTS

      Co_High : constant Co_Reading := 600;

    OPERATIONS
      NONE
    EXCEPTIONS
      NONE

  REQUIRED_INTERFACE
    OBJECTS

      Data_Logger

        TYPES
          NONE
        OPERATIONS

          Co_Log(Reading : Co_Reading);

        EXCEPTIONS
          NONE

      Operator_Console

        TYPES

          Alarm_Reason

        OPERATIONS

          Operator_Console.Alarm_Reason;

        EXCEPTIONS
          NONE

DATAFLOW

  Co_Reading => Data_Logger;
  Alarm_Reason => Operator_Console;

EXCEPTION_FLOW
  NONE

CYCLIC_OPERATION_CONTROL_STRUCTURE
  DESCRIPTION
    As there are no asynchronous transfers of control, there is no OBCS.
```

CONSTRAINED_OPERATIONS
NONE
HANDLED_EXCEPTIONS
NONE

INTERNALS
OBJECTS
NONE
TYPES
NONE
DATA

```
Co_Present : Co_Reading;
-- define control and status register
-- for the CO ADC
Cocsr : Device_Register.Csr;
for Cocsr'Address use 16#Aa1c#;
-- define the data register
Codbr : Co_Reading;
for Codbr'Address use 16#Aa1e#;
for Codbr'Size use One_Word;
Now : Duration;
```

OPERATIONS

```
Initialise
```

OPERATION_CONTROL_STRUCTURE
OPERATION `Initialise` IS
DESCRIPTION
The initialisation code
USED_OPERATIONS
NONE
PROPAGATED_EXCEPTIONS
NONE
HANDLED_EXCEPTIONS
NONE
CODE

```
-- enable device
Cocsr.Device := D_Enabled;
```

END OPERATION `Initialise`

OPERATION `Thread` IS
DESCRIPTION
The CO controller is a simplified version of the CH4 task. On finding a high level

its only role is to inform the operator.
USED_OPERATIONS

```
Operator_Console.Alarm;
Data_Logger.Co_Log
```

PROPAGATED_EXCEPTIONS
NONE
HANDLED_EXCEPTIONS
NONE
CODE

```
Cocsr.Operation := Set; -- start conversion
Now := Seconds(Clock);  -- wait for conversion
loop
   exit when Seconds(Clock) >= Now + 0.1;
end loop;

if not Cocsr.Done then
   Operator_Console.Alarm(Co_Device_Error);
else
   -- read device register for sensor value
   Co_Present := Codbr;
   if Co_Present > Co_High then
      Operator_Console.Alarm(High_Co);
   end if;
   Data_Logger.Co_Log(Co_Present);
end if;
```

END OPERATION Thread

CYCLIC_OPERATION_CONTROL_STRUCTURE
CODE
NONE
END_OBJECT Co_Sensor

9.4.12. Data Logger

OBJECT Data_Logger IS ACTIVE
DESCRIPTION
This object implements data logging and retrieval.

IMPLEMENTATION_CONSTRAINT
This is a non-terminal object.

PROVIDED_INTERFACE
TYPES
NONE
CONSTANTS

```
    NONE
  OPERATION_SETS
    NONE
  OPERATIONS

    procedure Co_Log(Reading : Co_Reading);
    procedure Ch4_Log(Reading : Ch4_Reading);
    procedure Air_Flow_Log(Reading : Air_Flow_Reading);
    procedure High_Low_Water_Log(Mark : Water_Mark);
    procedure Water_Flow_Log(Reading : Boolean);
    procedure Motor_Log(State : Motor_State_Changes);

  EXCEPTIONS
    NONE

REQUIRED_INTERFACE
  OBJECTS

    Environment_Monitor

      CONSTANTS
        NONE
      TYPES

        Ch4_Reading
        Co_Reading
        Air_Flow_Status

      OPERATIONS
        NONE
      EXCEPTIONS
        NONE

    Pump_Controller

      CONSTANTS
        NONE
      TYPES

        Motor_State_Changes

      OPERATIONS
        NONE
      EXCEPTIONS
        NONE
```

DATAFLOW

```
Motor_State_Change <= Pump_Controller;
Ch4_Reading <= Environment_Monitor;
Co_Reading <= Environment_Monitor;
Air_Flow_Reading <= Environment_Monitor;
```

EXCEPTION_FLOW
NONE

ACTIVE_OBJECT_CONTROL_STRUCTURE
DESCRIPTION
All operations are asynchronous
CONSTRAINED_OPERATIONS
`Co_Log` CONSTRAINED_BY ASER
`Ch4_Log` CONSTRAINED_BY ASER
`High_Low_Water_Log` CONSTRAINED_BY ASER
`Water_Flow_Log` CONSTRAINED_BY ASER
`Motor_Log` CONSTRAINED_BY ASER

INTERNALS
OBJECTS
not given here
TYPES
not given here
OPERATIONS
not given here

OPERATION_CONTROL_STRUCTURE
not given here

ACTIVE_OBJECT_CONTROL_STRUCTURE
IMPLEMENTED_BY
no given here

END_OBJECT `Data_Logger`

9.4.13. Operator Console

OBJECT `Operator_Console` IS ACTIVE

DESCRIPTION
This object implements the operator's console interface.

IMPLEMENTATION_CONSTRAINT
This is a non-terminal object.

PROVIDED_INTERFACE
TYPES

```
Alarm_Reason is (High_Methane, High_Co, No_Air_Flow,
                 Ch4_Device_Error, Co_Device_Error,
                 Pump_Dead, Unknown_Error);
```

CONSTANTS
NONE
OPERATION_SETS
NONE
OPERATIONS

```
procedure Alarm(Reason: Alarm_Reason)
```

EXCEPTIONS
NONE

REQUIRED_INTERFACE
OBJECTS

```
Pump_Controller
```

CONSTANTS
NONE
TYPES

```
Pump_Status
```

OPERATIONS

```
Request_Status
```

EXCEPTIONS

```
Not_Safe
```

DATAFLOW

```
Alarm_Reason <= Pump_Controller;
Alarm_Reason <= Environment_Monitor;
```

ACTIVE_OBJECT_CONTROL_STRUCTURE
DESCRIPTION
Operation is asynchronous
USED_OPERATIONS
not_given
EXCEPTIONS
not_given
CONSTRAINED_OPERATIONS
`Alarm` CONSTRAINED_BY ASER

INTERNALS
  OBJECTS
    not given here
  TYPES
    not given here
  OPERATIONS
    not given here
OPERATION_CONTROL_STRUCTURE
  not given here

ACTIVE_OBJECT_CONTROL_STRUCTURE
  IMPLEMENTED_BY
    no given here

END_OBJECT Operator_Console

9.5. Translation to Ada 95

Each of the objects shown in Figure 9.4 can potentially be implemented on a separate processor. However, for the purpose of this example we consider a single processor implementation.

Implementation for each of the terminal level objects can now be given. The decomposition appropriate to the pump controller was shown in Figure 9.5 and the high_low_water_sensor object is given in Figure 9.7. It is now possible to give the code for these objects.

9.5.1. Motor

```
with Device_Register_Types; use Device_Register_Types;
with System; use System;
with Motor_Rtatt; use Motor_Rtatt;
package Motor is -- PROTECTED

  type Pump_Status is (On, Off, Disabled);
  type Motor_State_Changes is (Motor_Started,
       Motor_Stopped, Motor_Safe, Motor_Unsafe);
  Pump_Not_Safe : exception;

  protected Obcs is
    pragma Priority(Motor_Rtatt.
                    Initial_Ceiling_Priority);
    procedure Not_Safe;
    procedure Is_Safe;
    function Request_Status return Pump_Status;
    procedure Set_Pump(To : Pump_Status);
```

```
  private
    Motor_Status : Pump_Status := Off;
    Return_Condition: Pump_Status := Off;
    -- define control and status register
    -- for the motor
    Pcsr : Device_Register_Types.Csr :=
      (Error_Bit => Clear, Operation => Set,
       Done => False, Interrupt => I_Enabled,
       Device => D_Enabled);
    for Pcsr'Address use 16#Aa14#;
  end Obcs;

  procedure Not_Safe renames Obcs.Not_Safe; -- ASER
  procedure Is_Safe renames Obcs.Is_Safe; -- ASER
  function Request_Status return Pump_Status renames
      Obcs.Request_Status; -- ASER
  procedure Set_Pump(To : Pump_Status) renames
      Obcs.Set_Pump; -- PSER
end Motor;

with Data_Logger;
with Ch4_Status; use Ch4_Status;
-- with Environment_Monitor;
package body Motor is

  protected body Obcs is
    procedure Not_Safe is
    begin
      -- Opcs_Not_Safe; generated inline
      if Motor_Status /= Disabled then
        Return_Condition := Motor_Status;
        Motor_Status := Disabled;
        Pcsr.Operation := Clear; -- turn off motor
        Data_Logger.Motor_Log(Motor_Unsafe);
      end if;
    end Not_Safe;
```

```
    procedure Is_Safe is
    begin
      -- Opcs_Is_Safe; generated inline
      Data_Logger.Motor_Log(Motor_Safe);
      if Return_Condition = On then
        Pcsr.Operation := Set; -- start motor
        Data_Logger.Motor_Log(Motor_Started);
        Motor_Status := Return_Condition;
      end if;
    end Is_Safe;

    function Request_Status return Pump_Status is
    begin
      -- Request_Status; generated inline
      return Motor_Status;
    end Request_Status;

    procedure Set_Pump(To : Pump_Status) is
    begin
      -- Opcs_Set_Pump ; generated inline
      if To = On then
        if Motor_Status = Off or
           Motor_Status = Disabled then
          if Ch4_Status.Read  = Motor_Safe then
            -- Environment_Monitor.Check_Safe
            Motor_Status := On;
            Pcsr.Operation := Set; -- turn on motor
            Data_Logger.Motor_Log(Motor_Started);
          elsif Motor_Status = Disabled then
            raise Pump_Not_Safe;
          end if;
        end if;
      else
        if Motor_Status = On then
          Pcsr.Operation := Clear; -- turn off motor
          Motor_Status:= Off;
          Data_Logger.Motor_Log(Motor_Stopped);
        end if;
      end if;
    end Set_Pump;
  end Obcs;
end Motor;
```

9.5.2. Water Flow Sensor Handling Object

```
package Flow_Sensor is -- CYCLIC
  type Water_Flow is (Yes, No);
                          -- calls Operator_Console.Alarm
                          -- calls Data_Logger.Water_Flow_Log
                          -- calls Motor.Request_Status
end Flow_Sensor;

with Ada.Real_Time; use Ada.Real_Time;
with Device_Register_Types; use Device_Register_Types;
with System; use System;
with Flow_Sensor_Rtatt; use Flow_Sensor_Rtatt;
with Ada_Real_Time.Cpu_Budgets;
use Ada.Real_Time.Cpu_Budgets;
with Motor; use Motor;
package body Flow_Sensor is

  -- periodic task
  Budget_Time : Time_Span :=
              Flow_Sensor_Rtatt.Thread.Budget;

  Flow : Water_Flow := No;
  Current_Pump_Status, Last_Pump_Status : Pump_Status;

  -- define control and status register
  -- for the flow switch
  Wfcsr : Device_Register_Types.Csr;
  for Wfcsr'Address use 16#Aa14#;

  procedure Opcs_Overrun_Of_Budget is separate;
    -- not given
  procedure Opcs_Periodic_Code is separate;

  procedure Opcs_Initialise is separate;
```

```
  task Thread;

  task body Thread is
    T: Time;
    Period : Time_Span := Flow_Sensor_Rtatt.Period;
  begin
    T:= Clock;
    Opcs_Initialise;
    loop
      Opcs_Periodic_Code;
      T := T + Period;
      delay until(T);
    end loop;
  end;

end Flow_Sensor;

separate(Flow_Sensor)
procedure Opcs_Initialise is
begin
  -- enable device
  Wfcsr.Device := D_Enabled;
end;

with Operator_Console; use Operator_Console;
with Data_Logger; use Data_Logger;
separate(Flow_Sensor)
procedure Opcs_Periodic_Code is
  My_Id : Budget_Id;
begin
  My_Id := Set_Budget(Budget_Time);
  Current_Pump_Status := Motor.Request_Status;
  if (Wfcsr.Operation = Set) then
    Flow :=  Yes;
  else
    Flow := No;
  end if;
  if Current_Pump_Status = On and
       Last_Pump_Status = On and Flow = No then
    Operator_Console.Alarm(Pump_Dead);
  end if;
  Last_Pump_Status := Current_Pump_Status;
  Data_Logger.Water_Flow_Log(Flow);
```

```
    Cancel(My_Id);
  exception
    when Budget_Overrun =>
      if My_Id = Last_Overrun then
        Opcs_Overrun_Of_Budget;
      else
        Cancel(My_Id);
        raise;
      end if;
  end Opcs_Periodic_Code;
```

9.5.3. The Controller Object

```
with Hlw_Controller_Rtatt; use Hlw_Controller_Rtatt;
with System; use System;
with Ada.Interrupts; use Ada.Interrupts;
with Ada.Interrupts.Names; use Ada.Interrupts.Names;
package Hlw_Controller is -- PROTECTED

  protected Obcs is
    pragma Priority(Hlw_Controller_Rtatt.
                     Initial_Ceiling_Priority);
    procedure Sensor_High_Ih;
    pragma Attach_Handler(Sensor_High_Ih, Waterh_Interrupt);
           -- assigns interrupt handler
    procedure Sensor_Low_Ih;
    pragma Attach_Handler(Sensor_Low_Ih, Waterl_Interrupt);
           -- assigns interrupt handler
  private
  end Obcs;

  procedure Sensor_High_Ih renames Obcs.Sensor_High_Ih;
    -- PSER
  procedure Sensor_Low_Ih renames Obcs.Sensor_Low_Ih;
    -- PSER

end Hlw_Controller;
```

```
with Hlw_Handler; use Hlw_Handler;
package body Hlw_Controller is

 protected body Obcs is
   procedure Sensor_High_Ih is
   begin
     Hlw_Handler.Start(High);
   end Sensor_High_Ih;

   procedure Sensor_Low_Ih is
   begin
     Hlw_Handler.Start(Low);
   end Sensor_Low_Ih;
  end Obcs;
end Hlw_Controller;
```

9.5.4. Handler

```
with Hlw_Handler_Rtatt; use Hlw_Handler_Rtatt;
with Ada.Real_Time; use Ada.Real_Time;
package Hlw_Handler is -- SPORADIC

  type Water_Mark is (High, Low);

  protected Obcs is
    pragma Priority(Hlw_Handler_Rtatt.
                Initial_Ceiling_Priority);
    -- for the Start operation
    procedure Start(Int : Water_Mark);
    entry Wait_Start(Int : out Water_Mark;
                 Start_Time : out Ada.Real_Time.Time;
                 Overrun : out Boolean);
  private
    Start_Open : Boolean := False;
    Freq_Overrun : Boolean := False;
    W : Water_Mark;
    T : Ada.Real_Time.Time;
  end Obcs;
  procedure Start(Int : Water_Mark) renames Obcs.Start;

end Hlw_Handler;
```

```
with Device_Register_Types; use Device_Register_Types;
with System; use System;
package body Hlw_Handler is

   -- define control and status registers
   -- for the high and low water switches
   Hwcsr : Device_Register_Types.Csr;
   for Hwcsraddress use 16#Aa10#;
   Lwcsr : Device_Register_Types.Csr;
   for Lwcsr'Address use 16#Aa12#;

   procedure Opcs_Sporadic_Frequency_Overrun is separate;
      -- not given
   procedure Opcs_Overrun_Of_Budget is separate;
      -- not given
   procedure Opcs_Initialise is separate;
   procedure Opcs_Thread is separate;
   procedure Opcs_Start(Int : Water_Mark) is separate;

   task Thread is
     pragma Priority(Hlw_Handler_Rtatt.
                     Current_Thread_Priority);
   end Thread;

   protected body Obcs is
     procedure Start(Int : Water_Mark) is
     begin
       W := Int;
       T := Ada.Real_Time.Clock;
            -- log time of invocation request
       if Start_Open then
         Freq_Overrun := True;
       else
         Start_Open := True;
       end if;
     end Start;
```

```
      entry Wait_Start(Int :out  Water_Mark;
            Start_Time : out Ada.Real_Time.Time;
            Overrun : out Boolean) when Start_Open is
      begin
        Int := W;
        Start_Time := T;
        Overrun := Freq_Overrun;
        Freq_Overrun := False;
        Start_Open := False;
      end Wait_Start;
    end Obcs;

    task body Thread is
      Int : Water_Mark;
      Start_Time : Ada.Real_Time.Time;
      Overrun : Boolean;
    begin
      Opcs_Initialise;
      loop

        Obcs.Wait_Start(Int, Start_Time, Overrun);
        if Overrun then
          Opcs_Sporadic_Frequency_Overrun;
        else
          Opcs_Start(Int);
        end if;
        delay until (Hlw_Handler_Rtatt.Mat + Start_Time);
      end loop;
    end Thread;
  end Hlw_Handler;
```

```
with Data_Logger; use Data_Logger;
with Motor; use Motor;
with Cpu_Budgets; use Cpu_Budgets;
separate(Hlw_Handler)
procedure Opcs_Start(Int : Water_Mark) is
  My_Id : Budget_Id;
begin
  My_Id := Set_Budget(Hlw_Handler_Rtatt.Thread.Budget);
  if Int = High then
    Motor.Set_Pump(On);
    Data_Logger.High_Low_Water_Log(High);
  else
    Motor.Set_Pump(Off);
    Data_Logger.High_Low_Water_Log(Low);
  end if;
  Cancel(My_Id);
exception
  when Budget_Overrun =>
    if My_Id = Last_Overrun then
      Opcs_Overrun_Of_Budget;
    else
      Cancel(My_Id);
      raise;
    end if;
end Opcs_Start;

separate(Hlw_Handler)
procedure Opcs_Initialise is
begin
  Hwcsr.Device :=D_Enabled;
  Hwcsr.Interrupt := I_Enabled;
  Lwcsr.Device :=D_Enabled;
  Lwcsr.Interrupt := I_Enabled;
end Opcs_Initialise;
```

9.5.5. Environment Monitoring

The decomposition of the environment monitor subsystem is shown in Figure 9.7. Note that the "check_safe" subprogram allows the pump controller to observe the current state of the methane level without blocking, via the CH4_status protected object. All other components are periodic activities.

9.5.6. CH4_status Object

```
with Ch4_Status_Rtatt; use Ch4_Status_Rtatt;
package Ch4_Status is -- PROTECTED
  type Methane_Status is (Motor_Safe, Motor_Unsafe);
  protected Obcs is
    pragma Priority(Ch4_Status_Rtatt.
                    Initial_Ceiling_Priority);
    procedure Write (Current_Status : Methane_Status);
    function Read return Methane_Status;
  private
    Environment_Status : Methane_Status := Motor_Unsafe;
  end Obcs;

  function Read return Methane_Status renames Obcs.Read;
     -- PSER
  procedure Write (Current_Status : Methane_Status)
             renames Obcs.Write; -- PSER
end Ch4_Status;

package body Ch4_Status is

  protected body Obcs is
    procedure Write (Current_Status : Methane_Status) is
    begin
      Environment_Status := Current_Status;
    end Write;

    function Read return Methane_Status is
    begin
      return Environment_Status;
    end Read;
  end Obcs;

end Ch4_Status;
```

9.5.7. CH4 Sensor Handling Object

```
package Ch4_Sensor is -- CYCLIC
  type Ch4_Reading is new Integer range 0 .. 1023;
  Ch4_High : constant Ch4_Reading := 400;
                      -- calls Motor.Is_Safe
                      -- calls Motor.Not_Safe
                      -- calls Operator_Console.Alarm
                      -- calls Data_Logger.Ch4_Log
end Ch4_Sensor;


with Ada.Real_Time; use Ada.Real_Time;
with System; use System;
with Ch4_Sensor_Rtatt; use Ch4_Sensor_Rtatt;
with Device_Register_Types; use Device_Register_Types;
with Ada.Real_Time.Cpu_Budgets;
use Ada.Real_Time.Cpu_Budgets;
package body Ch4_Sensor is

  Ch4_Present : Ch4_Reading;
  -- define control and status register
  -- for the CH4 ADC
  Ch4csr : Device_Register_Types.Csr;
  for Ch4csr'Address use 16#Aa18#;
  -- define the data register
  Ch4dbr : Ch4_Reading;
  for Ch4dbraddress use 16#Aa1a#;
  Jitter_Range : constant Ch4_Reading := 40;
  Now : Time;

  Budget_Time : Time_Span :=
              Ch4_Sensor_Rtatt.Thread.Budget;
  procedure Opcs_Overrun_Of_Budget is separate;
     -- not given
  procedure Opcs_Periodic_Code is separate;
  procedure Opcs_Initialise is separate;

  task Thread is
    pragma Priority(Ch4_Sensor_Rtatt.
                    Initial_Thread_Priority);
  end Thread;
```

```
  task body Thread is
    T: Time;
    Period : Time_Span := Ch4_Sensor_Rtatt.Period;
  begin
    T:= Clock;
    Opcs_Initialise;
    loop
      Opcs_Periodic_Code;
      T := T + Period;
      delay until(T);
    end loop;
  end;
end Ch4_Sensor;

separate(Ch4_Sensor)
procedure Opcs_Initialise is
begin
  -- enable device
  Ch4csr.Device := D_Enabled;
end Opcs_Initialise;

with Operator_Console; use Operator_Console;
with Motor; use Motor;
with Ch4_Status; use Ch4_Status;
with Data_Logger; use Data_Logger;
with Calendar; use Calendar;
separate(Ch4_Sensor)
procedure Opcs_Periodic_Code is
  My_Id : Budget_Id;

begin
  My_Id := Set_Budget(Budget_Time);
  begin
    Ch4csr.Operation := Set; -- start conversion
    Now := Clock;    -- wait for conversion
    loop
      exit when Clock >= Now + Milliseconds(100);
    end loop;
    if not Ch4csr.Done then
      Operator_Console.Alarm(Ch4_Device_Error);
```

```
    else
      -- read device register for sensor value
      Ch4_Present := Ch4dbr;
      if Ch4_Present > Ch4_High then
        if Ch4_Status.Read = Motor_Safe then
          Motor.Not_Safe; --pump_controller.not_safe
          Ch4_Status.Write(Motor_Unsafe);
          Operator_Console.Alarm(High_Methane);
        end if;
      elsif (Ch4_Present < (Ch4_High - Jitter_Range)) and
            (Ch4_Status.Read = Motor_Unsafe) then
        Motor.Is_Safe; -- Pump_Controller.Is_Safe
        Ch4_Status.Write(Motor_Safe);
      end if;
      Data_Logger.Ch4_Log(Ch4_Present);
    end if;

  exception
    when others =>
      -- One possible exception that would be caught is
      -- Constraint_Error which could theoretically be
      -- generated by the sensor reading going out of
      -- range. However, as this is a 10 bit device
      -- register which is mapped to the least
      -- significant end of a 16 bit memory location
      -- the error should not occur, and is therefore
      -- not caught explicitly.
      Operator_Console.Alarm(Unknown_Error);
      Motor.Not_Safe; -- Pump_Controller.Not_Safe
      Ch4_Status.Write(Motor_Unsafe);
      -- try and turn motor off before terminating
  end;
  Cancel(My_Id);

exception
  when Budget_Overrun =>
    if My_Id = Last_Overrun then
      Opcs_Overrun_Of_Budget;
    else
      Cancel(My_Id);
      raise;
    end if;

end Opcs_Periodic_Code;
```

9.5.8. Air Flow Sensor Handling Object

```
package Air_Flow_Sensor is -- CYCLIC
  type Air_Flow_Status is (Air_Flow, No_Air_Flow);
                       -- calls Data_Logger.Air_Flow_Log
                       -- calls Operator_Console.Alarm
end Air_Flow_Sensor;

with Device_Register_Types; use Device_Register_Types;
with System; use System;
with Ada.Real_Time; use Ada.Real_Time;
with Air_Flow_Sensor_Rtatt; use Air_Flow_Sensor_Rtatt;
with Cpu_Budgets; use Cpu_Budgets;
package body Air_Flow_Sensor is

  Air_Flow_Reading : Boolean;
  -- define control and status register
  -- for the flow switch
  Afcsr : Device_Register_Types.Csr;
  for Afcsr'Address use 16#Aa20#;

  Budget_Time : Time_Span :=
              Air_Flow_Sensor_Rtatt.Thread.Budget;
  procedure Opcs_Overrun_Of_Budget is separate;
      -- not shown
  procedure Opcs_Periodic_Code is separate;
  procedure Opcs_Initialise is separate;

  task Thread is
    pragma Priority(Air_Flow_Sensor_Rtatt.
                    Initial_Thread_Priority);
  end Thread;
```

```
  task body Thread is
    T: Time;
    Period : Time_Span := Air_Flow_Sensor_Rtatt.Period;
  begin
    T:= Clock;
    Opcs_Initialise;
    loop
      Opcs_Periodic_Code;
      T := T + Period;
      delay until(T);
    end loop;
  end Thread;
end Air_Flow_Sensor;

separate(Air_Flow_Sensor)
procedure Opcs_Initialise is
begin
  -- enable device
  Afcsr.Device := D_Enabled;
end Opcs_Initialise;

with Operator_Console; use Operator_Console;
with Data_Logger; use Data_Logger;
separate(Air_Flow_Sensor)
procedure Opcs_Periodic_Code is
  My_Id : Budget_Id;
begin
  My_Id := Set_Budget(Budget_Time);
  -- read device register for flow indication
  -- (operation bit set to 1);
  Air_Flow_Reading := Afcsr.Operation = Set;
  if not Air_Flow_Reading then
    Operator_Console.Alarm(No_Air_Flow);
    Data_Logger.Air_Flow_Log(No_Air_Flow);
  else
    Data_Logger.Air_Flow_Log(Air_Flow);
  end if;
  Cancel(My_Id);
```

```
exception
  when Budget_Overrun =>
    if My_Id = Last_Overrun then
      Opcs_Overrun_Of_Budget;
    else
      Cancel(My_Id);
      raise;
    end if;
end Opcs_Periodic_Code;
```

9.5.9. CO Sensor Handling Object

```
package Co_Sensor is   -- CYCLIC
  type Co_Reading is new Integer range 0 .. 1023;
  Co_High : constant Co_Reading := 600;
                         -- calls Data_Logger.Co_log
                         -- calls Operator_Console.Alarm
end Co_Sensor;


with Ada.Real_Time; use Ada.Real_Time;
with System; use System;
with Device_Register_Types; use Device_Register_Types;
with Co_Sensor_Rtatt; use Co_Sensor_Rtatt;
with Cpu_Budgets; use Cpu_Budgets;
package body Co_Sensor is

  Co_Present : Co_Reading;
  -- define control and status register
  -- for the CO ADC
  Cocsr : Device_Register_Types.Csr;
  for Cocsr'Addres use 16#Aa1c#;
  -- define the data register
  Codbr : Co_Reading;
  for Codbr'Address use 16#Aa1e#;
  Now : Time;
```

```
  Budget_Time : Time_Span :=
                Co_Sensor_Rtatt.Thread.Budget;
  procedure Opcs_Overrun_Of_Budget is separate;
            -- not shown
  procedure Opcs_Periodic_Code is separate;
  procedure Opcs_Initialise is separate;

  task Thread is
    pragma Priority(Co_Sensor_Rtatt.Initial_Thread_Priority);
  end Thread;

  task body Thread is
    T: Time;
    Period : Time_Span := Co_Sensor_Rtatt.Period;
  begin
    T:= Clock;
    Opcs_Initialise;
    loop
      Opcs_Periodic_Code;
      T := T + Period;
      delay until(T);
    end loop;
  end;
end Co_Sensor;

separate(Co_Sensor)
procedure Opcs_Initialise is
begin
  -- enable device
  Cocsr.Device := D_Enabled;
end Opcs_Initialise;
```

```
with Operator_Console; use Operator_Console;
with Data_Logger; use Data_Logger;
with Calendar; use Calendar;
separate(Co_Sensor)
procedure Opcs_Periodic_Code is
  My_Id : Budget_Id;
begin
  My_Id := Set_Budget(Budget_Time);
  Cocsr.Operation := Set; -- start conversion
  Now := Clock;  -- wait for conversion
  loop
    exit when Clock >= Now + Milliseconds(100);
  end loop;
  if not Cocsr.Done then
    Operator_Console.Alarm(Co_Device_Error);
  else
    -- read device register for sensor value
    Co_Present := Codbr;
    if Co_Present > Co_High then
      Operator_Console.Alarm(High_Co);
    end if;
    Data_Logger.Co_Log(Co_Present);
  end if;
  Cancel(My_Id);

exception
  when Budget_Overrun =>
    if My_Id = Last_Overrun then
      Opcs_Overrun_Of_Budget;
    else
      Cancel(My_Id);
      raise;
    end if;
end Opcs_Periodic_Code;
```

9.5.10. Data Logger

```
with Co_Sensor; use Co_Sensor;
with Ch4_Sensor; use Ch4_Sensor;
with Air_Flow_Sensor; use Air_Flow_Sensor;
with Hlw_Handler; use Hlw_Handler;
with Flow_Sensor; use Flow_Sensor;
with Motor; use Motor;
package Data_Logger is -- ACTIVE

  procedure Co_Log(Reading : Co_Reading); -- ASER
  procedure Ch4_Log(Reading : Ch4_Reading); -- ASER
  procedure Air_Flow_Log(Reading : Air_Flow_Status);-- ASER
  procedure High_Low_Water_Log(Mark : Water_Mark); -- ASER
  procedure Water_Flow_Log(Reading : Water_Flow); -- ASER
  procedure Motor_Log(State : Motor_State_Changes); -- ASER

end Data_Logger;
```

9.5.11. Operator Console

```
package operator_console is -- ACTIVE

  type alarm_reason is (high_methane, high_co, no_air_flow,
                        ch4_device_error, co_device_error,
                         pump_dead, unknown_error);
  procedure alarm(reason: alarm_reason); -- ASER

  -- calls Request_Status in Pump_Controller
  -- calls Set_Pump in Pump_Controller

end operator_console;
```

9.5.12. Real-Time Attributes

We assume that all objects only operate in one mode. Optimisations are therefore possible for the representation and access to the attributes.

```
with Rta; use Rta;
with System; use System;
with Ada.Real_Time; use Ada.Real_Time;
package Motor_Rtatt is

   -- for each operation
   Not_Safe : Operation_Attributes := (
              To_Time_Span(0.05),
              To_Time_Span(0.05));
   Is_Safe : Operation_Attributes := (
             To_Time_Span(0.05),
             To_Time_Span(0.05));
   Request_Status : Operation_Attributes := (
                    To_Time_Span(0.05),
                    To_Time_Span(0.05));
   Set_Pump : Operation_Attributes := (
              To_Time_Span(0.1),
              To_Time_Span(0.1));

   -- for PROTECTED objects
   Initial_Ceiling_Priority: constant := 6;

end Motor_Rtatt;

with Rta; use Rta;
with System; use System;
with Ada.Real_Time; use Ada.Real_Time;
package Flow_Sensor_Rtatt is

   -- for each CYCLIC and SPORADIC object
   Thread : Thread_Attributes := (2,
            To_Time_Span(0.0), Hard,
            To_Time_Span(3.0), To_Time_Span(0.15),
            To_Time_Span(0.15));

   -- for each CYCLIC object
   Period : Time_Span :=  To_Time_Span(60.0);
   Offset : Time_Span := To_Time_Span(0.0);
   Initial_Thread_Priority : constant Priority := 2;

end Flow_Sensor_Rtatt;
```

```
with Rta; use Rta;
with System; use System;
with Ada.Real_Time; use Ada.Real_Time;
package Hlw_Controller_Rtatt is

  -- for each operation
  Sensor_High_Ih : Operation_Attributes :=
                  (To_Time_Span(0.01), To_Time_Span(0.01));
  Sensor_Low_Ih : Operation_Attributes :=
                 (To_Time_Span(0.01), To_Time_Span(0.01));

  -- for PROTECTED objects
  Initial_Ceiling_Priority :  constant Priority :=  9;

end Hlw_Controller_Rtatt;

with Rta; use Rta;
with System; use System;
with Ada.Real_Time; use Ada.Real_Time;
package Hlw_Handler_Rtatt is

  -- for each CYCLIC and SPORADIC object
  Thread : Thread_Attributes := (1,
           To_Time_Span(0.0), Hard,
           To_Time_Span(20.0), To_Time_Span(0.1),
           To_Time_Span(0.1));

  -- for SPORADIC objects
  Initial_Ceiling_Priority :  constant Priority := 10;
  Initial_Thread_Priority :  constant Priority := 1;

  -- for each SPORADIC object
  Mat  :  Time_Span := To_Time_Span(100.0);
  Start : Operation_Attributes := (0.01, 0.01);

end Hlw_Handler_Rtatt;
```

```
with Rta; use Rta;
with System; use System;
with Ada.Real_Time; use Ada.Real_Time;
package Ch4_Status_Rtatt is

   -- for each operation
   Read : Operation_Attributes :=
          (To_Time_Span(0.01), To_Time_Span(0.01));

   -- for PROTECTED objects
   Initial_Ceiling_Priority :  constant Priority :=  7;

end Ch4_Status_Rtatt;

with Rta; use Rta;
with System; use System;
with Ada.Real_Time; use Ada.Real_Time;
package Ch4_Sensor_Rtatt is

   -- for each CYCLIC and SPORADIC object
   Thread : Thread_Attributes := (5,
            To_Time_Span(0.0), Hard,
            To_Time_Span(1.0), To_Time_Span(0.25),
            To_Time_Span(0.25));

   -- for each CYCLIC object
   Period : Time_Span := To_Time_Span(5.0);
   Offset : Time_Span := To_Time_Span(0.0);
   Initial_Thread_Priority : constant Priority := 5;

end Ch4_Sensor_Rtatt;
```

```
with Rta; use Rta;
with System; use System;
with Ada.Real_Time; use Ada.Real_Time;
package Co_Sensor_Rtatt is

  -- for each CYCLIC and SPORADIC object
  Thread : Thread_Attributes := (4,
           To_Time_Span(0.0), Hard,
           To_Time_Span(1.0), To_Time_Span(0.15),
           To_Time_Span(0.15));

  -- for each CYCLIC object
  Period : Time_Span := To_Time_Span(60.0);
  Offset : Time_Span  := To_Time_Span( 0.0);
  Initial_Thread_Priority : constant Priority := 4;

end Co_Sensor_Rtatt;

with Rta; use Rta;
with System; use System;
with Ada.Real_Time; use Ada.Real_Time;
package Air_Flow_Sensor_Rtatt is

  -- for each CYCLIC and SPORADIC object
  Thread : Thread_Attributes := (3,
           To_Time_Span(0.0), Hard,
           To_Time_Span(2.0), To_Time_Span(0.15),
           To_Time_Span(0.15));

  -- for each CYCLIC object
  Period : Time_Span :=  To_Time_Span(60.0);
  Offset : Time_Span  := To_Time_Span(0.0);
  Initial_Thread_Priority : constant Priority := 3;
end Air_Flow_Sensor_Rtatt;
```

9.6. Conclusion

This case study has illustrated the use of HRT-HOOD in the design of the mine control system. Ideally the process by which the logical and physical architectures were derived should be supported by CASE tools. In particular there is a need to check the consistency of the HRT-HOOD diagrams and the corresponding ODSs, and to integrate this support with a schedulability analyser and deadline simulator.

The case study has also illustrates the mapping that will occur to Ada 95.

10 The Olympus Attitude and Orbital Control System

This chapter describes the details of, and the experiences gained from, a case study undertaken in collaboration with the authors on the design and re-implementation of the Olympus Satellite's Attitude and Orbital Control Systems (AOCS). The goal of the study was to demonstrate that real-time systems can be implemented using Ada and its tasking facilities. The system was designed using HRT-HOOD, analysed using Deadline Monotonic Scheduling Analysis, and implemented on a M68020-based system using a modified Ada compiler and run-time support system (the modifications are compatible with those proposed for Ada 95). Our results indicate that systems can be designed to have the flexibility given by multi-tasking solutions, and yet still obtain the same levels of guarantees as those given by cyclic executives.

Much of the material used in this Chapter is a result of work undertaken by Chris Bailey. We gratefully acknowledge his contribution.

10.1. Background to the Case Study

Although Ada 83 has made some inroads into the real-time embedded computer systems market, often these systems are programmed in sequential Ada using cyclic executives. Over the last decade much research has been undertaken on the use of process-based systems using preemptive priority scheduling. Techniques such as Rate Monotonic[64] and Deadline Monotonic[63] schedulability analysis are now gaining favour; furthermore the real-time limitations of Ada 83 are well understood[20,22,73] and extensive changes have been made in Ada 95 to make the language more responsive to the needs of the real-time community.[79]

This chapter describes the details of, and the experiences gained from, a case study on the design and re-implementation of the Olympus Satellite's Attitude and Orbital Control Systems (AOCS). The goal of the study[11] was to demonstrate that real-time systems can be implemented using Ada and its tasking facilities (see Locke [65] for a discussion on the advantages of process-based scheduling over cyclic executive). The chapter is structured as follows:

- An overview of the system, giving the functional and non-functional requirements.
- The design of the system using the HRT-HOOD design method.
- The implementation of the design in Ada 83 running on top of an augmented stand-alone Ada run-time support system kernel. The kernel has been augmented by

those facilities which will be available in Ada 95.

- A discussion on the problems encountered and how they were solved.

10.2. The Modelled System: The Olympus AOCS

The Olympus satellite was launched in July 1989 as the world's largest and most powerful civil three-axis-stabilised communications satellite. Situated at longitude 19 degrees West, Olympus provides direct broadcast TV and 'distance learning' experiments to Italy and Northern Europe.

The AOCS subsystem exists to acquire and maintain the desired spacecraft position and orientation. The AOCS software may operate in six modes of operation, of these the Normal Mode is the most complex and is used for the greatest percentage of the satellites lifetime. It is this mode that the study primarily modelled.

Hardware Architecture

As depicted in Figure 10.1, the Normal Mode software is embedded in the Spacecraft Microcomputer Module (SMM) and communicates with the following devices over a serial data bus

- A Telemetry & Telecommand Subsystem (TMTC),
- An InfraRed Earth Sensor (IRES),
- A Digital Sun Sensor (DSS),
- A Rate Gyro Sensor (RGS),
- Four reaction wheels (RWs),
- Thrusters.

There is a cold-standby SMM which is powered up should a failure be detected. In this case study we are interested in real-time aspects of the system, and therefore we shall assume the system's hardware and software is reliable.

Serial bus messages are placed on the bus according to the priority of the transmitting device. Gyros have the highest priority, followed by the TMTC and the SMMs. The bus has time slots reserved for replies, ensuring that SMM requests for sensor data receive responses within a 960 μs time slot. There is a 10ms real-time clock.

Software Requirements

The object of 'Normal Mode' attitude control is to maintain the satellite's orientation to Earth. This is to be achieved by a 200 ms cyclic task that forces IRES roll and pitch angles and a rate-gyro derived yaw angle to zero by controlling the speed of four reaction wheels.

For two spells per day of 15 minutes, a one second period task calibrates the gyro drift rate by comparing the yaw estimate, derived from an integration of gyro rate, against the sun and earth positions. The gyro angle and gyro drift rate are corrected at the end of each spell. The gyro data is received approximately every 100ms (without being requested).

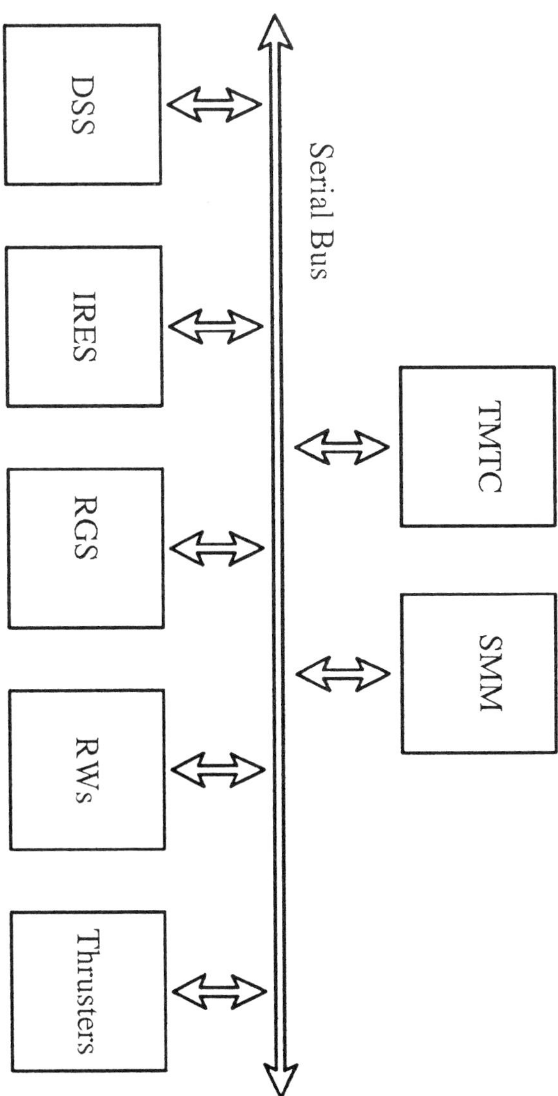

Figure 10.1: The AOCS Hardware Architecture

In addition to the regular activities identified above, further attitude control functions occur at less frequent or irregular intervals.

- Momentum dumping is triggered when the speed of any reaction wheel exceeds a preset threshold. This consists of a reduction in the reaction wheel speed in a series of steps, while compensating bursts of thruster firings prevent the loss of Earth-pointing. Dumping on the three axes operates independently.

- The Telemetry and Telecommand Subsystem (TMTC) routinely requests status information from the AOCS software. This task, which has a minimum period of 62.5 ms, keeps the ground informed of the spacecraft state. The SMM is unable to respond to a telemetry request in the same bus time slot, so it transmits a response in a later time slot.
- A telecommand function allows ground to enable or disable control, to enable or disable dumping, to trigger gyro calibration, or to set a reaction wheel failed or operational. Telecommands can occur at a minimum interval of 190 ms.

For a full description of the Requirements see Bailey.[10]

The application selected contains many typical features of embedded real-time space software,[9] namely:

- Cyclic tasks,
- Sporadic tasks,
- Hard real-time tasks,
- Soft real-time tasks,
- Background tasks,
- Communication over a bus.

The operational software was coded in 9989 assembler and scheduled by a cyclic scheduler.

10.3. The Logical Architecture Design

Figure 10.2 shows the context in which the control software is to be designed.

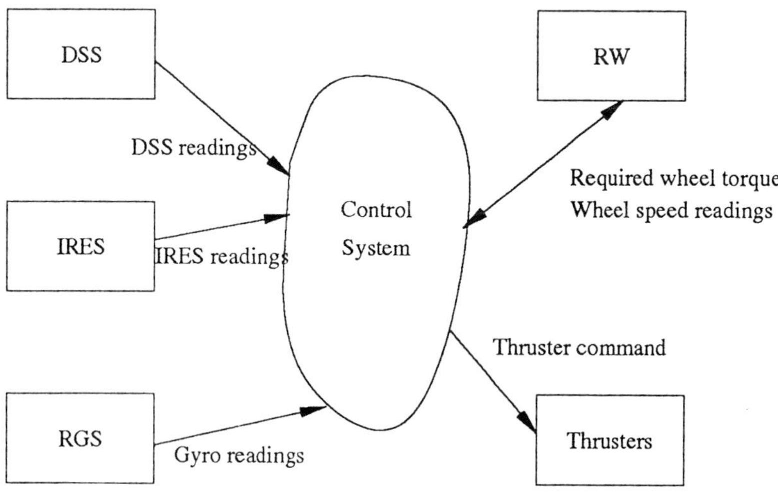

Figure 10.2: Relationship between External Devices and the Control System

It should be noted that all sensor readings and actuator commands are communicated over a serial bus.

10.3.1. First Level Decomposition

Figure 10.3 shows the first level decomposition of the software. The software is basically constructed from three subsystems: the CONTROLLER, an active object, which implements the main AOCS software (monitoring sensors, initiating actuators, and implementing the control laws), the interface to the Telemetry and Telecommand Subsystem (which is shown for convenience as two objects: a terminal sporadic object to implements the incoming commands, and an active object which is responsible for sending status reading to ground), and a bus controller subsystems (which again for convenience is shown as two objects: an active object for handling incoming messages, and a terminal protected objects for placing data in the hardware FIFO buffer for output onto the network). The diagram also shows the real-time attributes of the terminal objects which have been added.

10.3.2. The CONTROLLER Object

The decomposition of the controller object is shown in Figure 10.4. It consists of:

- several protected objects — which are used to control access to data which is shared between the activities of the system; in particular the SERIAL BUS IP protected object is used to encapsulate the data received from the bus (via the RECEIVE FROM BUS object),
- the "CONTROL_LAW" terminal cyclic object — which implements the basic control laws, and is therefore responsible for maintaining the Satellites Earth point.
- the "SENSOR" active object — — which monitors the satellites sensors and provide information for the CONTROL_LAW object,
- the "ACTUATORS" active object — which controls access to the actuators.

Further decomposition of the SENSOR object is shown in Figures 10.5 to 9.

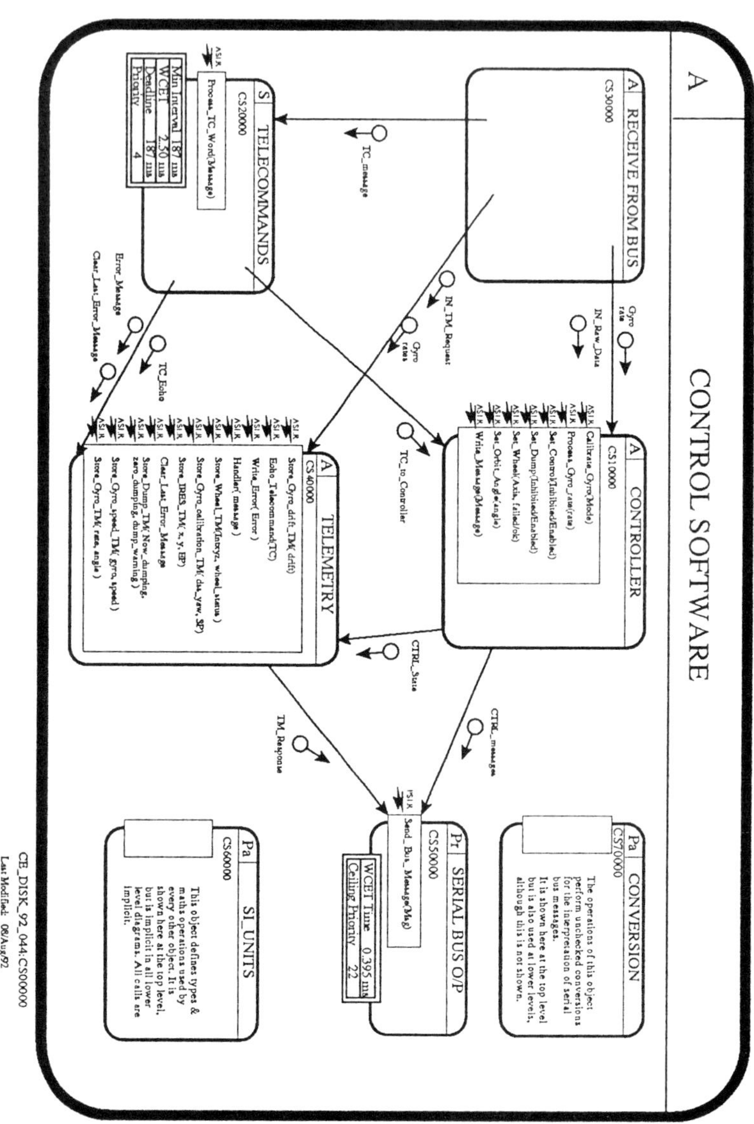

Figure 10.3: The Control Software

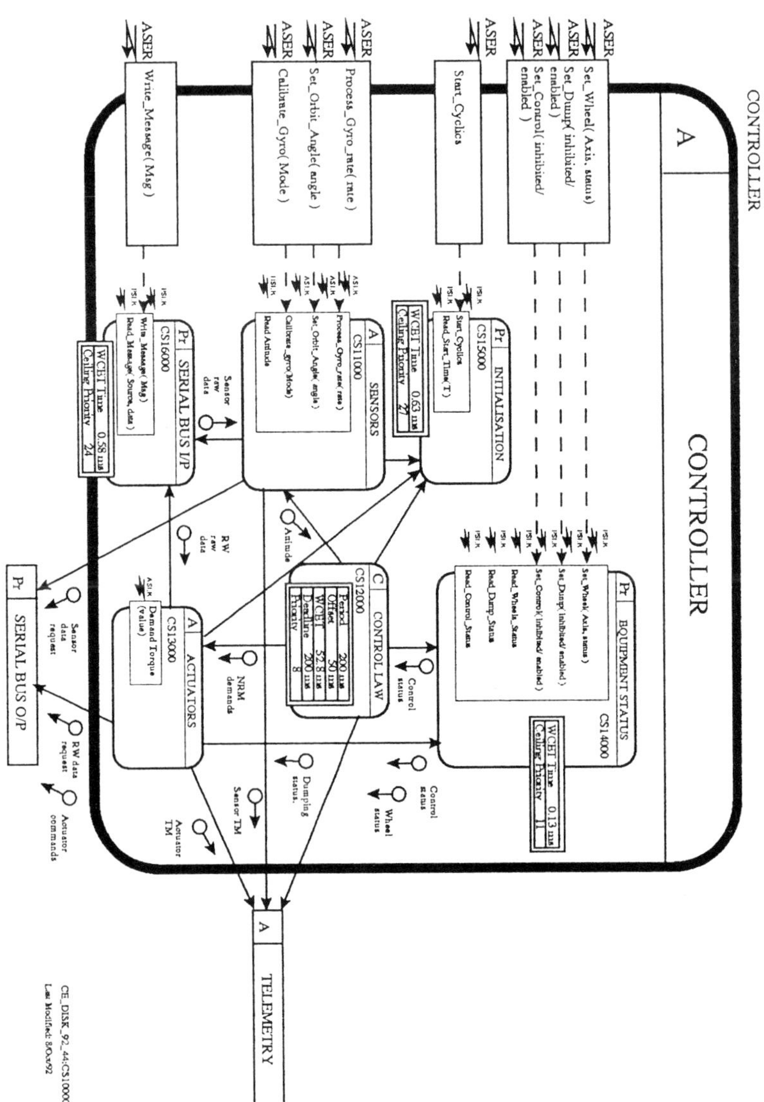

Figure 10.4: The CONTROLLER Object

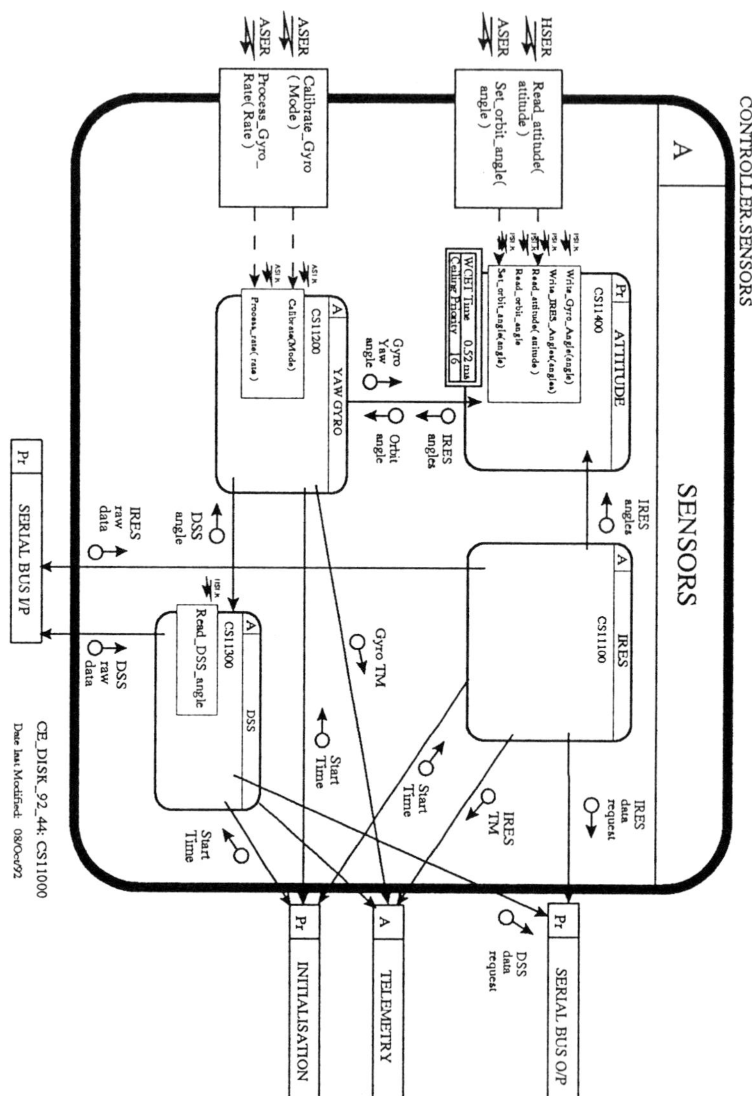

Figure 10.5: The SENSOR Object

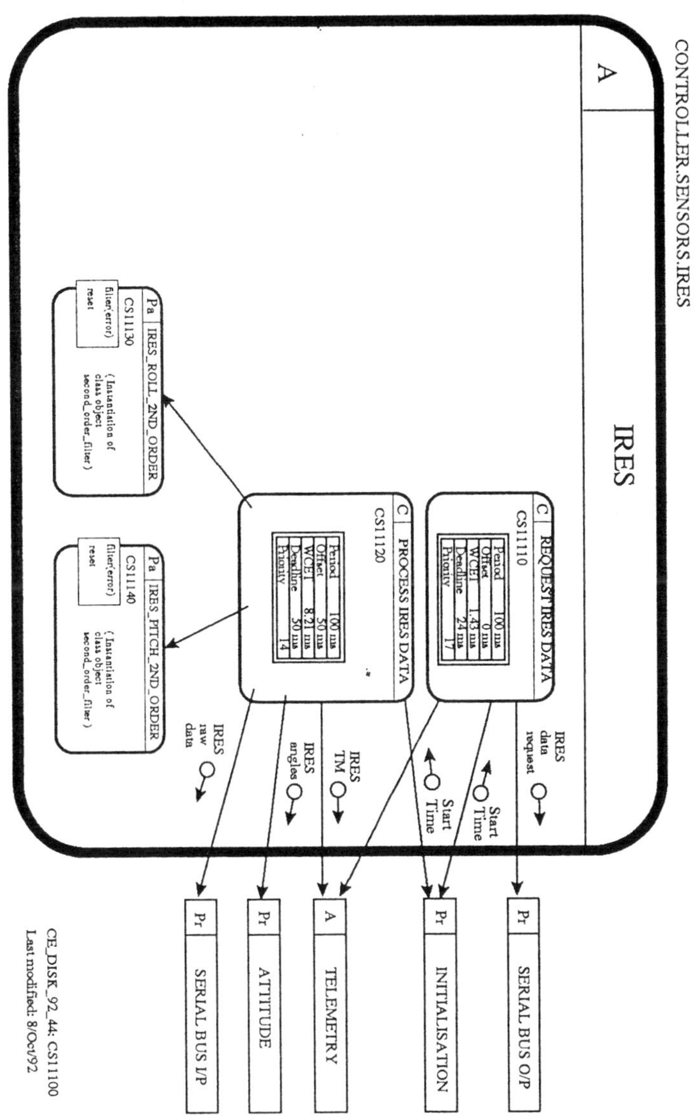

Figure 10.6: The IRES Object

The IRES sensor controller consists of two precedent constrained cyclic objects with a time offset between the two implementing the required synchronisation. The first object, REQUEST IRES DATA, sends a request to the IRES (via the serial bus). The second object will receive and interpret the sensor values. The relative time offset between the task releases and the deadline of the first object ensures that the sensor device has a chance to respond (at least 30 ms).

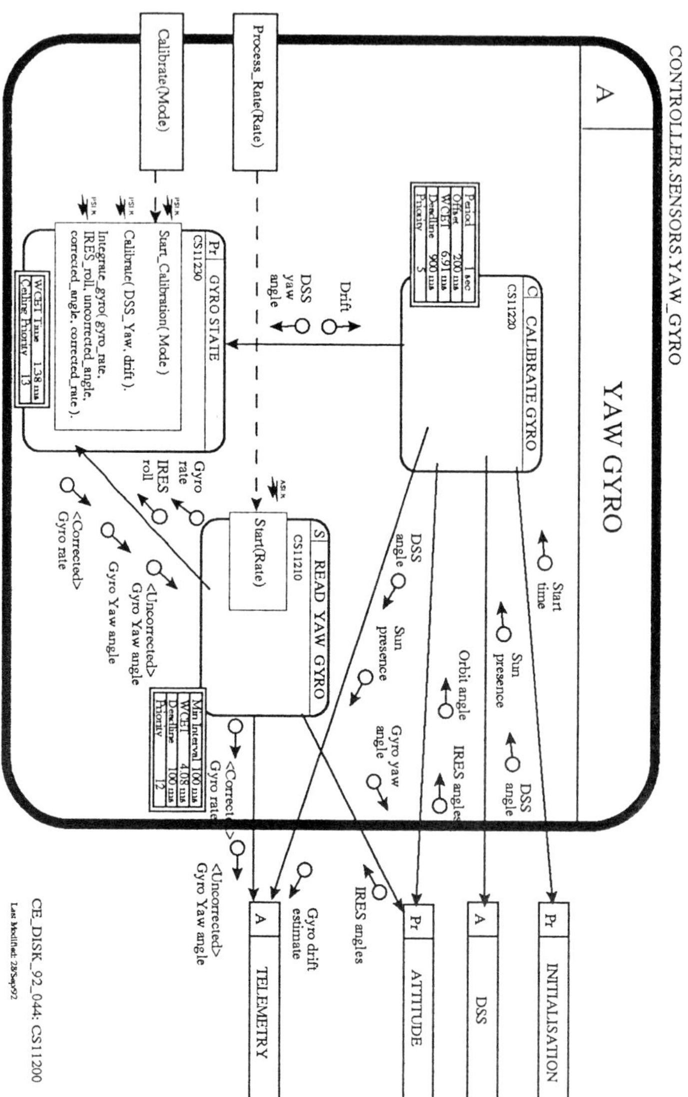

Figure 10.7: The YAW GYRO Object

The gyro processing consists of cyclic object which calibrates the gyros every second. The sporadic object, READ YAW GYRO, processes the incoming data from the sensors. Note that this object is a sporadic even though the data comes in regularly. This is because the sensor has a different clock and there may be some jitter on the received data.

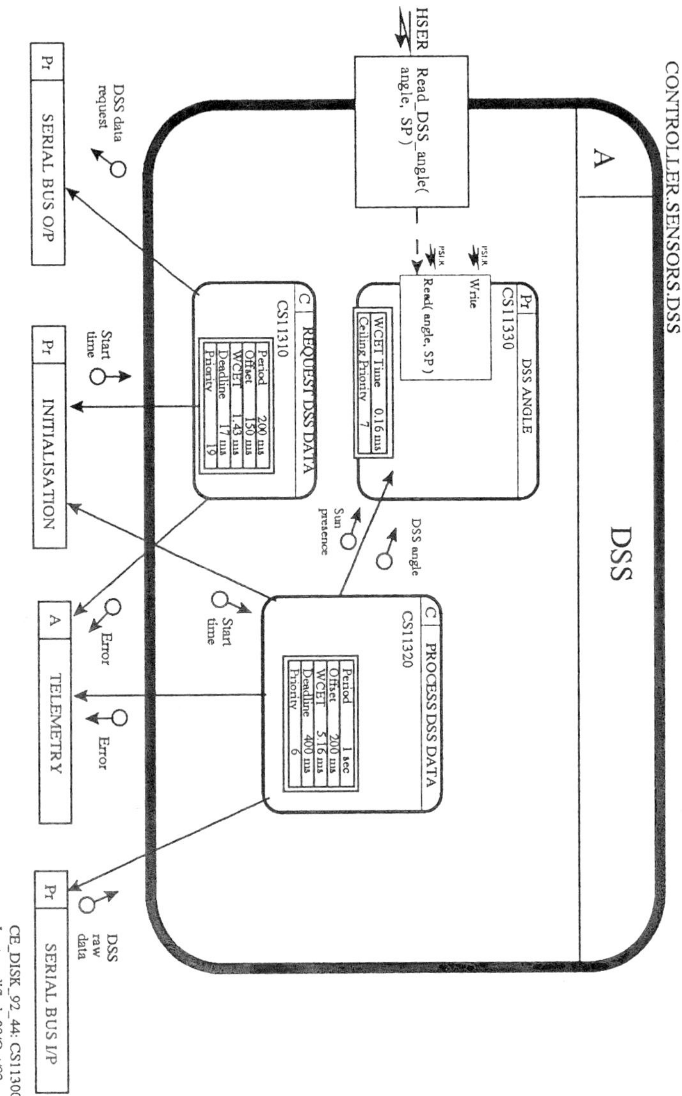

Figure 10.8: The DSS Object

The digital sun sensor, again, consists of two cyclic objects with a relative offset between them. The deadline of 20ms on the REQUEST DATA object ensures that there are 30 ms available for delivery of the message and for the sensor to respond.

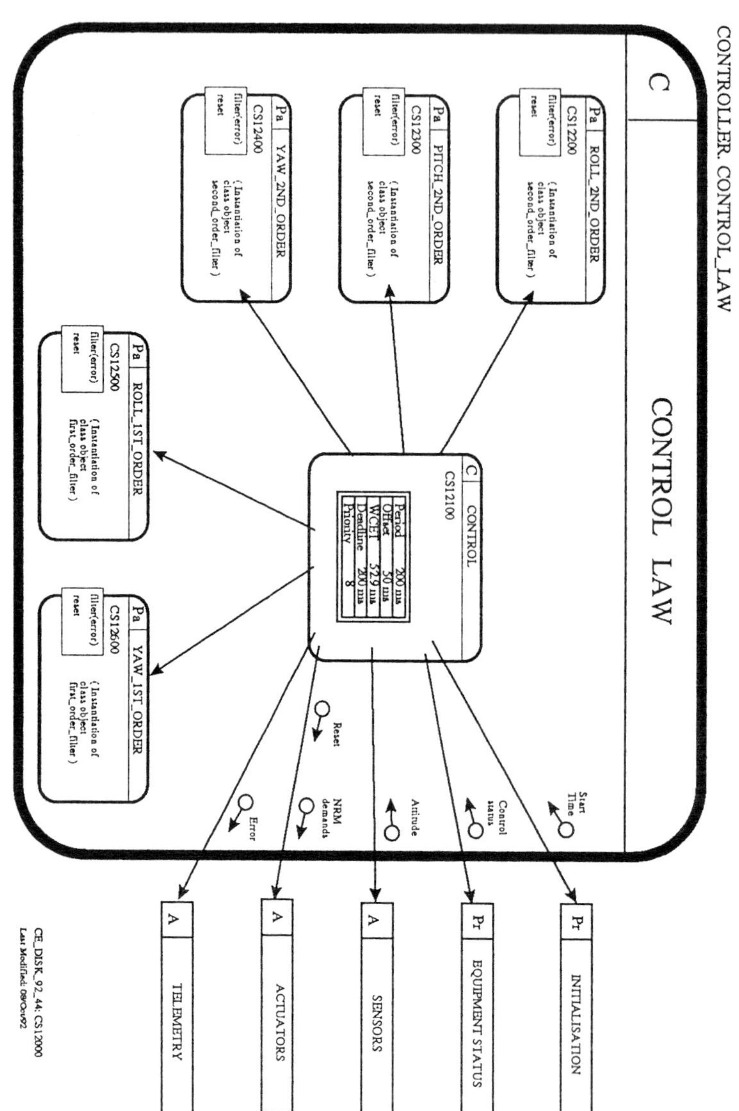

Figure 10.9: The CONTROL LAW Object

The CONTROL LAW Object consists of a 200 ms cyclic object which implements the basic control laws for the satellite by monitoring the sensors and sending commands to the actuators.

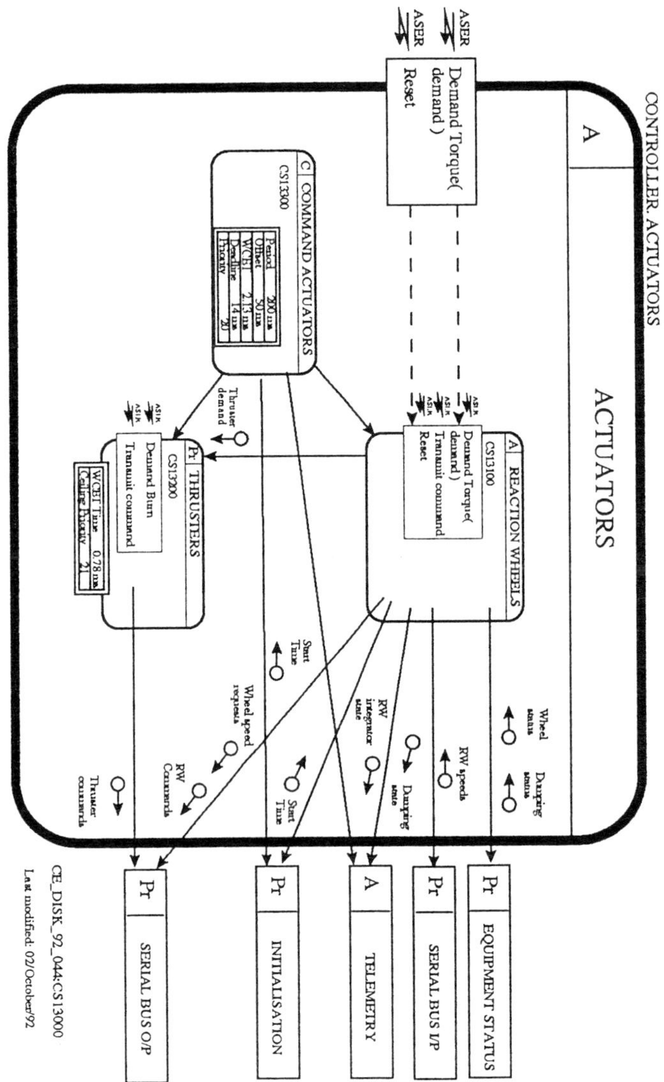

Figure 10.10: The ACTUATORS Object

The ACTUATOR object encapsulates the reaction wheels and the thruster actuators. The decomposition of the REACTION WHEELS object is shown in Figures 10.11 to 10.13.

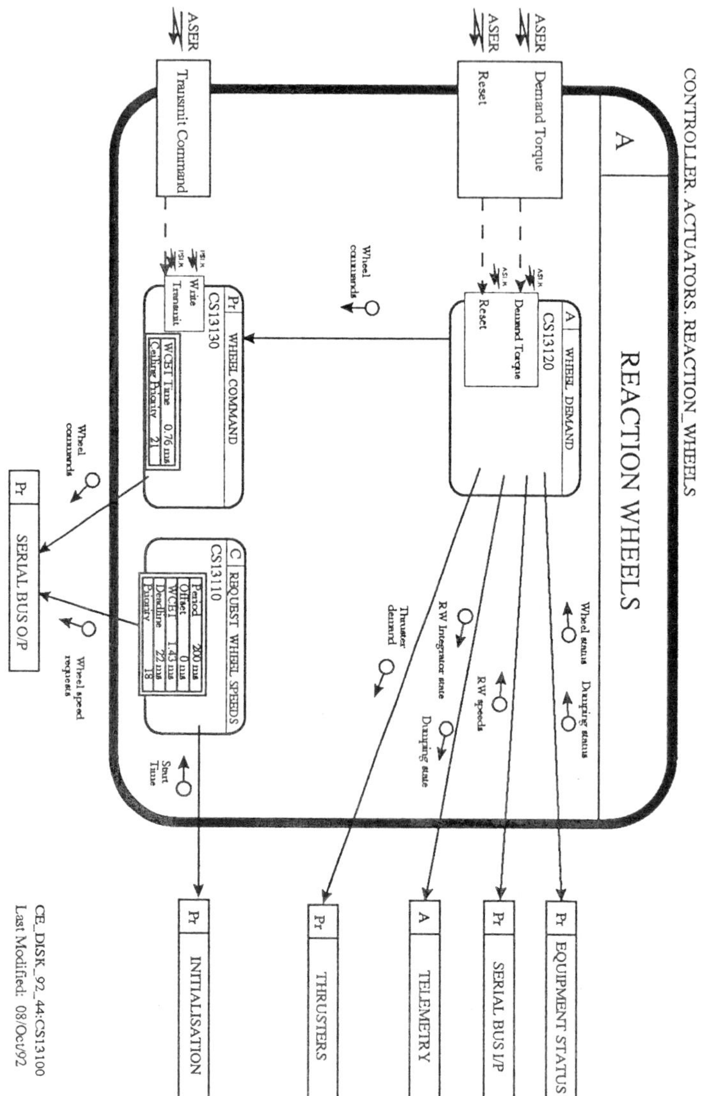

Figure 10.11: The REACTION WHEELS Object

The REACTION WHEELS object controls the operation of the reaction wheels. Although we consider the wheels to be actuators, they also provide readings giving their current speed. The cyclic object REQUEST WHEEL SPEEDS requests those speeds every 200 ms. The values returned from the device is held in the SERIAL BUS I/P Object.

The WHEEL COMMAND object simply constructs the command for forwarding onto the SERIAL BUS I/O object for transmission across the network.

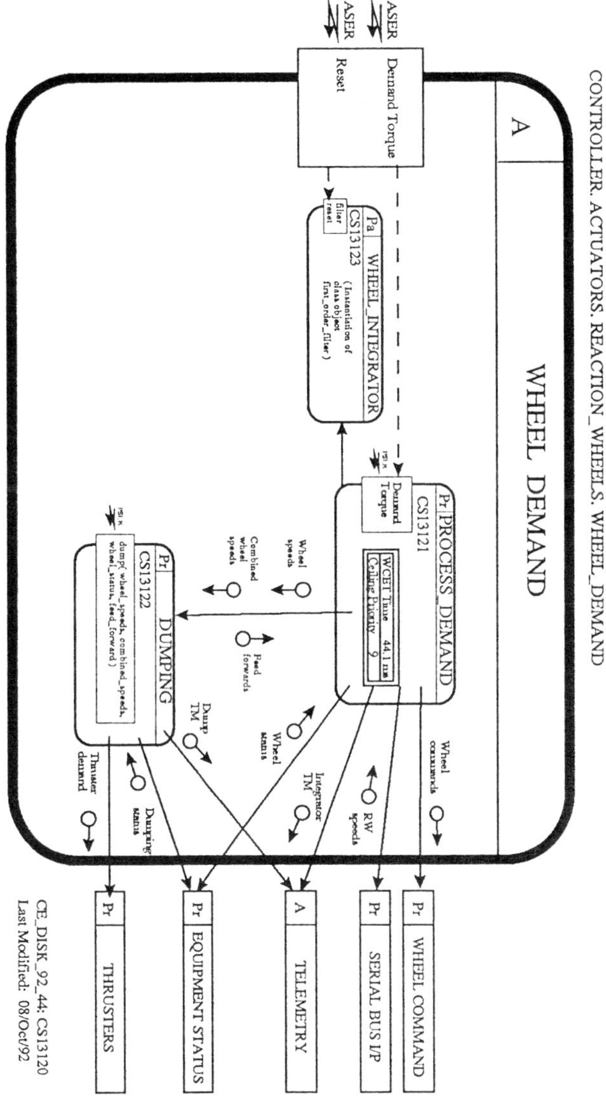

Figure 10.12: The WHEEL DEMAND Object

The WHEEL DEMAND object provides the functionality for driving the actuator. It consists of two protected object which either issue commands to the wheels or instructs the thrusters to fire.

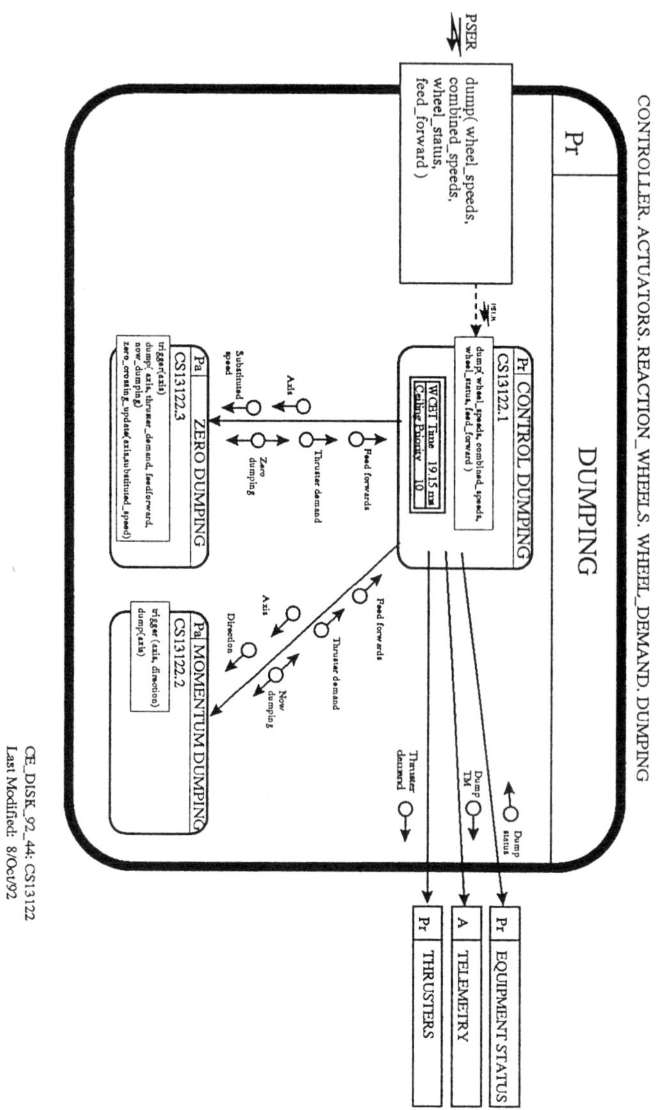

Figure 10.13: The DUMPING Object

10.3.3. RECEIVE FROM BUS Object

The RECEIVE FROM BUS object handles the input data arriving on the serial bus. A sporadic object responds to the bus interrupt and places the data into a buffer. A cyclic objects then retrieves the data and passes it on to the appropriate receiving object.

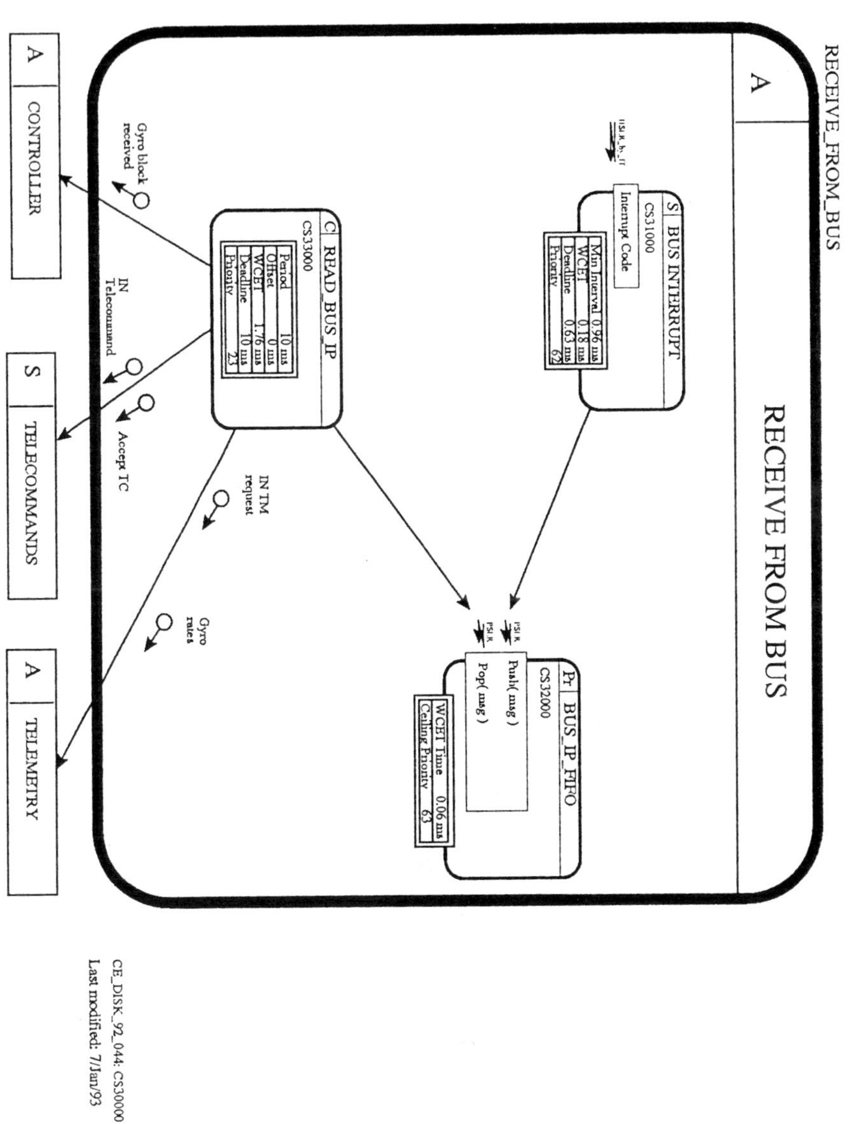

Figure 10.14: RECEIVE FROM BUS Object

10.3.4. TELEMETRY Object

The TELEMETRY object is responsible for storing sensor readings and passing them to ground when requested.

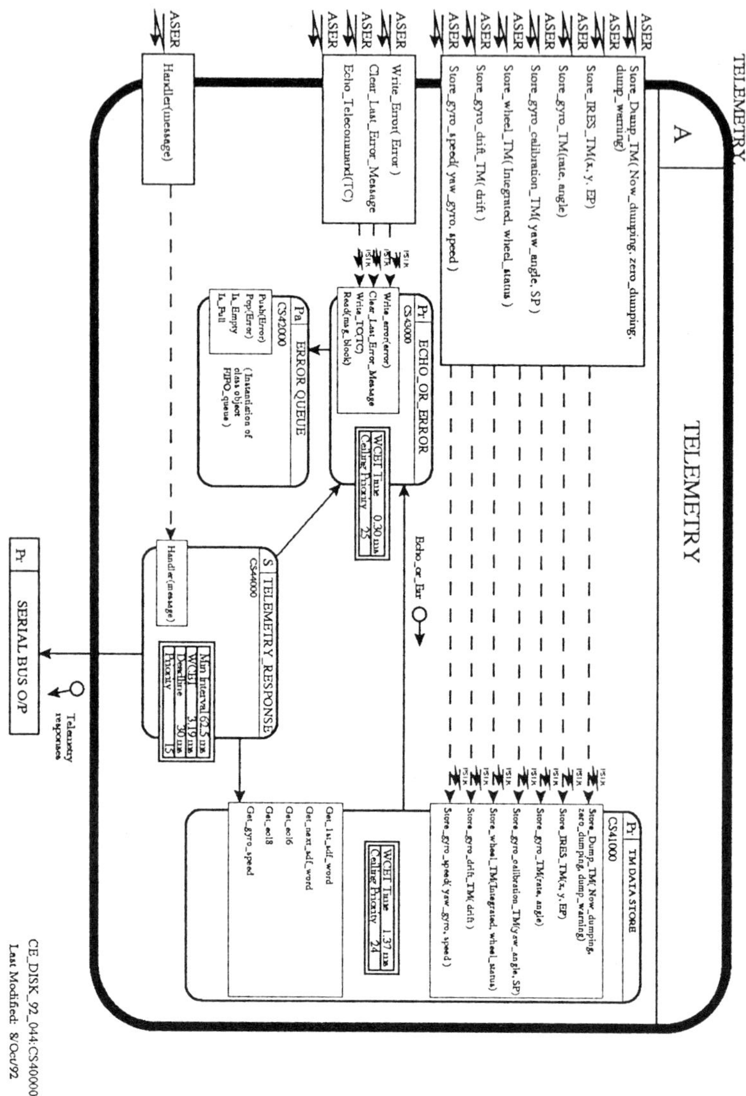

Figure 10.15: TELEMETRY Object

10.4. The Physical Architecture Design

This section describes the proposed execution environment for the system and how the software design is mapped to it.

10.4.1. The Hardware Platform

The case study runs on 2 VME boards containing a 68020 processor, a 68881 floating point coprocessor, 1 MByte RAM, timers, and dedicated chips for communication over the Olympus serial bus.

These new cards replace the current Spacecraft Microcomputer Module (SMM) in the Olympus [AOCS] Engineering Model testbed to demonstrate successful operation of the unit.

For ease of timing analysis, neither the processor cache nor DMA was used.

10.4.2. The Operating System

The YSE Ada compiler[1] and a modified stand-alone run-time kernel has been used in this study. The system has been modified to support some of the new features proposed for the new Ada 95 standard.

- Large priority range
- Priority queuing
- Protected objects — implemented as optimised passive tasks
- Delay until

For ease of analysis, the following features of Ada have not been used:

- Dynamic task creation and abortion,
- Access types,
- Dynamic memory allocation or deallocation,
- The Ada83 rendezvous (rather protected objects are used).

10.4.3. Mapping the Software Architecture to the Execution Environment

The final HRT-HOOD design containing the following application terminal objects:

- 9 cyclic objects,
- 3 sporadic objects,
- 14 protected objects
- 16 passive objects,

and produced a total of 3300 lines of Ada code.

Criticality

The requirement for the AOCS identified two levels of software criticality. The majority of software must be guaranteed to met all deadlines; these are denoted as HARD. The remaining software is non-critical and denoted as SOFT. In the following scheduling analysis, the HARD process are considered first. All SOFT processes are assigned priorities below those used by the HARD.

Schedulability Analysis

Using Deadline Monotonic scheduling analysis[6, 63] each task is given a unique priority (*P*); the higher the priority the shorter the task's deadline. In order to undertake schedulability analysis it is necessary to have the real-time characteristics of the terminal objects. The worst case execution times of the objects were calculated using a tool constructed for the case study.[43] Other characteristics, such as cycle times of periodic processes are known from the requirements. The following tables summarise the real-time characteristic of the final system, and gives the task and protected object priorities which were calculated by another tool.[26] All times are in ms. A more detailed description of the results is given in Burns et al.[28]

Task name	Importance	Period	Offset	WCET	Required Deadline	Achieved Deadline	Priority
RTC	HARD	50	0	0.28	9.0	3.52	27
Read_Bus_IP	HARD	10	0	1.76	10.0	6.99	23
Command_Actuators	HARD	200	50	2.13	14.0	13.52	20
Request_DSS_Data	HARD	200	150	1.43	17.0	15.87	19
Request_Wheel_Speeds	HARD	200	0	1.43	22.0	18.22	18
Request_IRES_data	HARD	100	0	1.43	24.0	23.37	17
Process_IRES_data	HARD	100	50	8.21	50.0	44.13	14
Control_Law	HARD	200	50	52.84	200.0	183.50	8
Process_DSS_Data	HARD	1000	200	5.16	400.0	198.38	6
Calibrate_Gyro	HARD	1000	200	6.91	900.0	389.49	5

Table 10.1: Cyclic Thread Characteristics

Task name	Importance	Minimum Arrival Time	WCET	Required Deadline	Achieved Deadline	Priority
Bus_Interrupt	INTERRUPT	0.96	0.18	0.63	--	62
Telemetry_Response	HARD	62.5	3.19	30.0	28.73	15
Read_Yaw_Gyro	HARD	100.0	4.08	100.0	55.84	12
Telecommands	SOFT	187.0	2.5	187.0	FAIL	4

Table 10.2: Sporadic Thread Characteristics

Note that to implement sporadic objects requires a synchronisation agent. This is in fact a form of protected object. There are three of these identified as X_OBCS.

10.5. Problems Encountered

The main problem we encountered was associated with handling the interrupts off the bus. Originally we attempted to map a sporadic object to the bus interrupt and to have this object pass on the data to the various sensor objects. A sporadic object in HRT-

PR name	WCET	Ceiling Priority
Bus_IP_FIFO	0.06	63
Initialisation	0.63	27
Serial_Bus_IP	0.58	24
Read_Yaw_Gyro.OBCS	0.15	24
Telecommands.OBCS	0.15	24
Telemetry_Response.OBCS	0.15	24
Echo_or_Error	0.30	25
Serial_Bus_OP	0.39	22
Wheel_command	0.76	21
Thrusters	0.78	21
TM_data_store	1.37	24
Attitude	0.52	16
Gyro_state	1.38	13
Equipment_status	0.13	11
Control_dumping	19.15	10
Process_demand	44.09	9
DSS_angle	0.16	7

Table 10.3: Protected Tasks Characteristics

HOOD is mapped to an Ada task and a passive (protected) Ada task (representing an Ada 95 protected object). The passive task handles the interrupt, and releases the other task to deal with the received data. However, the schedulability analysis indicated that with a minimum inter-arrival time of 960 μs for the sporadic (the estimated minimum time between interrupts) the system was not schedulable. In fact the overhead of entering the Ada passive task and releasing the sporadic was almost 960 μs. This reflected the prototype nature of our modifications to the Ada compiler to implement the equivalent of an Ada 95 protected record.

The problem was overcome by

1) Modifying the analysis — it was recognised that the minimum inter-arrival rate for the interrupt was not sustained over a long period; the system (for analysis purposes) was more accurately modelled as the sum of four interrupts (representing the four different message types that could be received from the sensors — called Message_here, TM_here, Z1_here, and TC_here in Appendix 2), each with their characteristic minimum interarrival time. These task are not themselves analysed but rather are used to model the interference on other tasks.

2) Modifying the design — the system design was modified to that presented in this book; the interrupt handling sporadic simply places the data in a buffer and a cyclic object removes the data at an even rate (calculated to ensure that all the sensor objects get adequately fresh data) and calls the relevant sensor objects.

3) Modifying the translation to Ada 83 — the interrupt handling was implemented as a call to an Ada procedure (thus representing an optimised protected object in Ada 95).

Given these modifications, the analysis indicated that the one soft task, TELECOMMANDS, would not meet its deadline in the worst case. In practice, the task did meet its deadline because of pessimism in:

1) the analysis techniques — some of the objects have offsets specified relative to other objects; the equations we were using assumed all tasks had a critical instance when this is clearly not the case (we now have more sophisticated analysis which will handle task offsets[82]).

2) the kernel — in order to take into account the overheads introduced by the kernel, we had to make some assumptions in the analysis techniques; these, on closer inspection, were a little pessimistic.[30]

3) the WCET tool — we estimate that our worst case execution time tool[43] is between 5-15% pessimistic because the tool does not model the m68020 internal pipeline and because some hardware times are data dependent and the tool has to assume the worst case.

10.6. Summary

The goal of the project was to illustrate that hard real-time systems can be programmed in a multi-tasking Ada environment, and yet give the same guarantees as those offered by the cyclic executive approach. The following points should be emphasised:

1) The use of a multi-tasking design introduced flexibility into the design; for example when the early design was shown not to meet its deadline it was not necessary to redo complex cyclic schedules. Instead the design could be easily altered and the schedulability analysis re-done.

2) The use of deadline monotonic scheduling, together with offsets between processes allowed input and output jitter to be kept to a minimum.

3) It is extremely important to model accurately the performance properties of the real-time operating system kernel if the scheduling analysis is to be relied on.

11 Conclusions

HRT-HOOD was developed in response to a requirement to demonstrate the use of Ada for hard real-time systems development for the European Space Agency. Initially our intention was to generate a set of guidelines on how to use Ada for hard real-time system development and that these should form the basis of an informal design method. However, it soon became clear that these guidelines would be used by ESA and their contractors in conjunction with structured design methods such as HOOD or Mascot. We were concerned that although structured design methods do allowed hard real-time systems to be engineered, they also allow many other systems to be constructed whose timing properties can not be analysed off-line. We therefore attempted to incorporate our guidelines into HOOD so that all systems designed according to the new method would have analysable timing properties.

HRT-HOOD provides support for an approach to real-time systems development which is based on the schedulability analysis of a group of concurrent activities; rather than, say, the cyclic execution of sequential program. Hence any attempt to evaluate the method inevitably also evaluates the effectiveness of the overall philosophy.

HRT-HOOD has been evaluated by several case studies undertaken by the European Space Agency. The first of these has been discussed in detail in Chapter 10 of this book. The study successfully showed that it is feasible and practical to use a process-based solution to on-board software systems, and to achieve the same level of control as that using the cyclic executive approach. The study highlighted that the positive benefits of using the HRT-HOOD method includes:[12, 83]

- a sound engineering approach with a mathematical basis
- a mathematical proof of schedulability resulting in a reduction in test duration; testing can be heavily targeted at logic checking with reduced emphasis on checking worst case loading situations
- flexible run-time scheduling allows changes to be made in the application structure without the costly redevelopment of cyclic schedules[65]
- tool support enables the techniques to be used by engineers rather than academic theoreticians

A similar study to the Olympus AOCS was the re-implementation of the Data Handling Software for the EURECA Platform. Again, positive results have been reported.[44, 83] Particularly encouraging is that HRT-HOOD was reported as aiding the construction of reusable components.

Although tool support for schedulability analysis and worst case execution time analysis exists, as of yet there is no tool support for the HRT-HOOD design process and Ada translation. However, work is underway on upgrading of the Intecs Sistemi 3.1 HOOD toolset to support HRT-HOOD object types and the translation process.

Outstanding Issues and Future Developments

The technology for implementing hard real-time systems on single processors is now relatively mature. However, supporting methods are few and far between. HRT-HOOD is one of the first structured design method which seriously attempts to address the issue of designing hard real-time systems. It is therefore inevitable that the method will need refinement as further experience with its use is accrued. Furthermore there are several outstanding HOOD-related issues which need more consideration. These are:

1) The exact role of *environment* objects in hard real-time system design. The current proposals do allow hard real-time objects to use *environment* objects. We assume that these objects are *root* objects and that the real-time characteristics of their *terminal* objects have been defined.

2) The issue of guaranteeing behaviour during mode changes is difficult. For example, consider a cyclic *thread* which has finished executing one cycle and is waiting for the next to begin. During this *period* it receives an asynchronous transfer of control request. To wake the *thread* before its next cycle will invalidate the schedulability analysis. However, to wait until its next cycle will decrease response time. The current HRT-HOOD approach is to wait until the next cycle. A similar situation occurs with sporadic *threads*.

   The issue of timely mode changes requires further investigation.[81]

As well as being subject to further case studies, HRT-HOOD will be developed in a number of ways. Attention will be given to the issue of reuse and how real-time objects can be modified and re-applied with the minimum of re-work and re-analysis. Also, now Ada 95 is finalised the mapping to that language will be consolidated. In addition mappings to other languages will be investigated.

The great benefit that a structured design method provides is the explicit representation of structure. Formal methods, although they have obvious advantages in terms of their unambiguous definition of behaviour, do not adequately capture the software architecture of a design. Current work on HRT-HOOD involves using its present definition to give a framework for capturing structure and undertaking system-wide timing analysis but using a formal method to define the behaviour of individual objects (and/or their defined operations). Is is not envisaged that a formal method would be needed for all objects in a design but only those that are sufficiently complex or which are particularly safety critical and hence hazardous.

Appendix A: Terminology

The terminology given in this book is that used in HRT-HOOD. Many of the terms are also used in HOOD, and have similar meaning.

Active object

an object which controls the times at which requested invocations of its operations are executed, and which may spontaneously invoke operations in other objects

Budget time

the time allocated to an operation or a *thread* for its execution

Ceiling priority

a priority assigned to all *protected* objects; this priority is no lower than the maximum priority of all the *threads* that can call the constrained operations

Class

an object template which represents a reusable object with type and data parameters

Constrained operation

an operation of an *active*, *protected*, *cyclic*, or *sporadic* object whose execution is controlled by the status of the object (as defined in the OBCS)

Control flow

control flow between objects is represented by the *use* relationship; control flow within an object is defined by the OBCS

Cyclic object

an object which represents a periodic activity, it may spontaneously invoke operations in other objects, but the only operations it has itself are requests which demand immediate attention (they represent asynchronous transfer of control requests)

Data flow

flow of data via parameters to operations between used and using objects

Deadline

the time by which an operation or *thread* must finish its execution

Design process tree

the tree of objects of the system being designed, consisting of the root object and its successive decomposition into child objects until terminal objects are reached

Environment object

an object which represents the provided interface of another object used by the system to be designed, but which is not part of the Design Process Tree

Exception flow

an exception is an abnormal return from a used operation to a using object

Execution transformation

an attribute of a *cyclic* or *sporadic* object which indicates that it must be transformed so that the *thread* execution has extra delays introduced

Importance

assigned to each *cyclic* and *sporadic* object; represents whether the thread is a hard real-time thread or a soft real-time thread

Include relationship

expresses that an object is fully decomposed into a set of child objects that collectively provide the same functionality as the parent

Instance

an object that is created for a design from a class object

Internal operation

an operation that is defined in a terminal object to support the implementation of provided operations

Non-terminal object

an object which is decomposed into child objects

Object Description Skeleton - ODS

the formal notation of the design of an object

Object Control Structure - OBCS

part of the ODS that defines the control flow between constrained operations

Offset

an attribute of a *cyclic* object which indicates how long the *thread* must delay after creation before starting its cyclic operations

Operation Control Structure - OPCS

part of the ODS that defines the logic of the operation (external or internal) in terms of pseudocode

Op_control object

a HOOD object that implements the mapping between one parent operation and more than one child object operation

Operation_Set

an operation which stands for a list of operations in order to ease writing of long lists of operations

Passive object

an object which does not have control over when invocations of its operations are executed, and does not spontaneously invoke operations in other objects

Precedence constraints

the constraints placed on a parent *cyclic* or *sporadic* object that its children objects must have constraints placed on their order of execution

Period

how often the *thread* in a *cyclic* object executes

Priority

the priority given to *threads*, according to deadline monotonic scheduling theory

Protected object

an object which may control when requested invocations of its operations are executed, and does not spontaneously invoke operations in other objects; in general *protected* objects may *not* have arbitrary synchronisation constraints and must be analysable for their blocking times

Provided interface

part of the ODS that defines the interface of the object

Real-time attributes

attributes which are given to objects to describe their real-time characteristics

Required interface

part of the ODS that defines the interfaces of used objects

Root object

the top level object which represents the system to be designed

Sporadic object

an object which represent a sporadic activity, it may spontaneously invoke operations in other objects; each sporadic has a single operation which is called to invoke the thread; and one or more operations which are requests demanding immediate attention (they represent asynchronous transfer of control requests)

Terminal object

an object which is not decomposed into child objects

Thread

the internal concurrent activity of a *cyclic, sporadic* or *active* object

Use relationship

an object is said to use another object if the former requires one or more of the operations provided by the latter

Worst case execution time

the maximum amount of computation time that an operation or *thread* requires to execute

Appendix B: HRT-HOOD Definition Rules

B.1 Design Checking, Scoping and HRT-HOOD Rules

A HRT-HOOD design is a representation of a system to be designed which starts with a description of a parent *root* object with respect to its environment, and is refined into successive descriptions of child objects. In order to ensure consistency of these successive representations, HRT-HOOD defines rules, which shall be enforced at each parent-child decomposition step.

B.1.1 Scoping

Each object has a well defined scope consisting of

- the *provided interface* of its parent
- its own *provided interface*
- its own *required interface.*

For a *root* object the scope is reduced to:

- its own *provided interface*
- its own *required interface.*

B.1.2 HRT-HOOD rules

Within an object's scope, formal rules can be defined for checking consistency and completeness of a design description. An overview of the categories of rules is given now. A detailed description of each category follows.

[Category G] General definition rules give the basic definitions and properties of HRT-HOOD object types (*active*, *passive*, *protected*, *cyclic*, *sporadic*, *environment*, and *class* in the design process tree.

[Category U] Use relationship rules define the way an object can use another one within their respective scopes and according to their object type.

[Category I] Include relationship rules define the way a given parent object is implemented by a set of child objects according to their object types and components (Operation, Types, Exception, OBCS and Data-flows).

Decomposition rules describe the Implemented_by relationship properties according to the type of the associated operation (Constrained, Unconstrained, provided, required, internal, implemented_by, operation_set) and the status of the associated object (*terminal, non terminal*).

[Category O] Operation rules define the basic definitions and properties of the HRT-HOOD operations.

[Category V] Visibility and naming rules.

The scope of an HRT-HOOD entity is the scope of the object where it has been defined. In order to identify unambiguously entities where several objects' scopes overlap the following rules are defined:

— an object name is unique in the design process tree

— HRT-HOOD entities of a design shall be defined using the dot notation stating the object name where the entity has been declared

— the scope of a provided operation is the object's scope where it is declared, whereas the scope of an *internal operation* is the object internals

— types which can be manipulated by an object are as follows:

  — types provided by the object's Parent

  — types declared in its *provided interface*

  — types declared as internal types

  — types referenced in its *required interface*

— the scope of an exception declaration is the object's scope

[Category C] Consistency rules check consistency of child description with respect to their parent description:

— parent *real-time attributes* with respect to the union of the child real-time attributes

— parent required operations and exceptions with respect to the union of the child required operations and exceptions

— parent provided operations and exceptions with respect to the union of the child provided operations and exceptions.

Internal consistency checks ensure consistency of the HRT-HOOD entities within the ODS, and between the ODS and the diagrammatic description. These checks can be performed with one or more objects within the design process tree.

Further consistency rules may be added to address levels of software integrity.

B.2 General Definitions

G-1 An object which does not have a parent is called a *root* object.

G-2 A HRT-HOOD system design shall have only one *root* object but may use other *root* objects as *environment* objects.

G-3 An object may be of type *active*, *passive*, *protected*, *cyclic*, *sporadic*, *environment*, or *class*.

G-4 A *class* object may be *active*, *passive*, *protected*, *cyclic*, or *sporadic*.

G-5 *Active*, *passive*, *protected*, *cyclic*, and *sporadic* objects are mutually exclusive (an object can be of only one type).

G-6 A HRT-HOOD design may contain one or more *environment* object, each of which defines the interface of an object external to the system being designed.

G-7 A *class* object shall be defined with parameters that may be types or data.

G-8 The parameters of an *instance* of a *class* object shall correspond to the parameters of the class definition.

G-9 A *terminal* object which is *active*, *protected*, *cyclic*, or *sporadic* shall have at most one OBCS.

G-10 A *terminal* object which is *cyclic*, or *sporadic* shall have one thread.

G-11 A thread is a run-time schedulable entity.

G-12 Only *cyclic*, *sporadic*, *protected*, and *passive* objects may be configured into a hard real-time subsystem.

B.3 Use Relationship

U-1 An object may *use* another object, i.e. an object requires an operation of the used object.

U-2 An object shall not *use* itself.

U-3 *Passive* and *protected* objects shall not *use* each other in a cycle.

U-4 *Passive* objects shall not *use constrained operations* of an *active*, *protected*, *cyclic*, or *sporadic* object.

U-5 *Cyclic* or *sporadic* objects shall not *use constrained operations* of a *terminal active* object unless they are asynchronous. They may *use constrained operations* of a parent *active* object as long as the operations are implemented by child *protected* objects.

U-6 *Cyclic* or *sporadic* objects may *use* the *constrained operations* of other *cyclic* or *sporadic* objects.

U-7 *Cyclic* or *sporadic* objects may *use* the *constrained operations* on *protected* objects.

U-8 *Protected* objects may not *use constrained operations* of an *active*, *cyclic*, or *sporadic* object unless they are asynchronous.

U-9 If an object *uses* another object and has children, then one of its children shall *use* the other object.

U-10 A child object can *use* an uncle object only if its parent *uses* the uncle object.

U-11 An exception shall be raised only between two objects when one object *uses* the other object.

U-12 Two objects which have *precedence constraints* must have at least one *protected* object which they both *use*.

Other relationships may be added to support integrity levels.

B.4 Include Relationships

I-1 An object may *include* another object, i.e. an object may decompose into a set of objects which together provide the same functionality as the object.

I-2 An object that is not decomposed is called the *terminal* object.

I-3 An object that has an internal object is called a *non-terminal* object.

I-4 An object that is *included* in another object is called a *child* object.

I-5 An object shall not have more than one parent object.

I-6 *Passive* objects may have only *passive* children.

I-7 An object shall not decompose into or be decomposed from itself.

I-8 A *protected* object can be decomposed into one *protected* object and several *passive* objects.

I-9 A *cyclic* object can be decomposed into at least one *cyclic* object and several *passive*, *protected* and *sporadic* objects.

I-10 A *sporadic* object can be decomposed into at least one *sporadic* object and several *passive*, *cyclic* and *protected* objects.

I-11 An *active* object may decompose into any other object.

I-12 The OBCS for a *non-terminal* object is implemented by one or more child OBCSs.

I-13 The thread for a *non-terminal* object is implemented by one or more children threads.

I-14 The OBCS for a *terminal* object shall contain pseudo-code or source code (eg Ada)

I-15 All types, constants, exceptions, operations, *operation sets* provided by a *non-terminal* object must be implemented by a child object.

Other relationships may be added to support integrity levels.

B.5 Operations

O-1 An operation may be declared in the *provided interface* to an object or internal to a *terminal* object.

O-2 Each operation shall be provided by one and only one object.

O-3 Each provided and internal operation shall have an OPCS.

O-4 Each parent operation shall be implemented by a single operation of a child.

O-5 An operation of a child object may implement at most one parent operation.

O-6 A *constrained operation* of an *active* parent shall only be implemented by a *constrained operation* of an *active, protected, cyclic*, or *sporadic* child.

O-7 An operation of a *passive* parent shall only be implemented by a *passive* child object.

O-8 Constrained operations may only be declared by *active*, *protected*, *cyclic*, or *sporadic* objects.

O-9 An operation shall not be constrained if it is provided by a *passive* object.

O-10 An operation of a *terminal* object shall not be implemented by an operation of another object, but may call an internal operation in its OPCS.

O-11 An *active*, *cyclic*, *sporadic* or *protected* object to which an exception is raised must have an exception handler in the OPCS.

O-12 An Operation_set of a *non-terminal* object may be decomposed into operations and operation sets.

O-13 An Operation_set shall not have parameters.

O-14 An Operation_set of a *terminal* object shall be decomposed into individual operations.

O-15 An object shall not have both internal objects and internal operations (i.e. if one provided operation is implemented by an internal object, then all provided shall be implemented by internal objects).

O-16 If a parent *cyclic* or *sporadic* object has been decomposed then ALL its operations must be implemented by its children objects. (The parent has no run-time existence: i.e. no run-time OBCS or thread.)

O-17 A *cyclic* thread must not block waiting for one of its operation to be called.

O-18 A *cyclic* and *sporadic* object must have at least one internal operation. This is the operation which is called at each invocation of the thread.

O-19 A *cyclic* and *sporadic* object shall have an internal operation which is called if the deadline of the thread is missed (this represents the application level exception handler).

O-20 A *cyclic* and *sporadic* object shall have an internal operation which is called if the *budget time* of the thread is exceeded.

O-21 A *cyclic, passive* and *sporadic* object shall have internal operations which are called if the *budget time* of their provided operations are exceeded.

O-22 A *cyclic* object may only have un*constrained operation*s or constrained operations which cause an asynchronous transfer of control in the *cyclic*'s thread.

O-23 A *sporadic* object may only have un*constrained operations*, a single START operation, or constrained operations which cause an asynchronous transfer of control in the *sporadic*'s thread.

O-24 A *protected* object must not block once it begins execution.

B.6 Visibility

V-1 Each object name shall be unique within a design.

V-2 Each operation name shall be unique within an object.

V-3 An object or operation name must not be a HRT-HOOD reserved word; furthermore the names will be restricted by the reserved words in the target language.

B.7 Consistency

C-1 If an object A has a required exception or operation of an object B, then object A must use the object B, and object B must be a required object of object A in the ODS.

C-2 The parts of a HRT-HOOD design must be consistent

(a) the entities in the diagram must all be in the ODS:

- ODS for each object
- provided operations and exceptions
- exception handlers for each exception raised
- required object for each uncle object
- internal object for each child object
- data-flow for each data-flow drawn
- exception-flow for each exception-flow drawn
- *constrained operation* for each operation of an *active*, *cyclic*, *sporadic* and *protected* object
- OBCS for each *active*, *cyclic*, *sporadic* and *protected* object which has *constrained operations*
- threads for each *cyclic* and *sporadic* object
- OPCS for each provided operation

(b) consistency within the ODS:

- object names for required types, operations and exceptions must be in the list of objects (including *class* and *environment* objects) of the Required Interface.
- OPCS for each internal operation

C-3 Each parent diagram shall have an implemented_by link for each operation to a child operation.

C-4 The Provided Interface of a used object shall correspond to the Required Interface of the using object.

C-5 The *real-time attributes* of a parent object shall be consistent with the *real-time attributes* of its children.

— The *period* of each parent *cyclic* object must equal the *periods* of its *cyclic* children objects, and must equal the minimum arrival time of its *sporadic* children. However, a *sporadic* child may suffer from release jitter.[6]

— The children of a *cyclic* object must have *precedence constraints*. Furthermore, there must exist one *cyclic* child which only has precedence constraints of the AFTER type.

— The minimum arrival time of a parent *sporadic* object must equal the minimum arrival time of its children.

Appendix C: Object Description Skeleton (ODS) Syntax Summary

In this appendix a description of the HRT-HOOD ODS is given in a variant of Backus-Naur-Form. Much of this description is similar to HOOD, it has been given here for completeness. Note, however, that this description conforms to the style presented in the Ada 83 Language Reference Manual. The only difference being that ODS keywords are presented in upper case so that they are distinguished from Ada keywords (which are presented in bolded lower case in the Ada Reference Manual). A syntactic category prefixed by "Ada_" denotes a category which is defined in the Ada 83 Reference Manual.

Informal text can be placed anywhere. It takes the form of an Ada comment and has the same representation (i.e. starts with two adjacent hyphens and extends up to the end of the line). Where informal text is required by the syntax it will be indicated by the syntactic category "informal_text".

C.1 General Declarations

```
object_name_list ::=
  object_name
  {, object_name }

object_name ::=
  name

operation_name ::=
  name

operation_set_name ::=
  name

type_name ::=
  name
```

```
data_name ::=
   name

exception_name ::=
   name

node_name ::=
   name

mode_name ::=
   name

name ::=
   Ada_identifier

object_description ::=
   DESCRIPTION
      informal_text
```

C.2 Object ODS Structure

```
HRT_ods ::= simple_ods | class_ods | instance_ods | environment_ods

simple_ods ::=
   OBJECT object_name IS PASSIVE | ACTIVE | PROTECTED |
                 CYCLIC | SPORADIC

      passive_visible_part | active_visible_part |
      protected_visible_part | cyclic_visible_part |
      sporadic_visible_part


                  passive_hidden_part | active_hidden_part |
                  protected_hidden_part | cyclic_hidden_part |
                  sporadic_hidden_part

              END_OBJECT object_name
```

```
class_ods ::=
   OBJECT object_name IS CLASS PASSIVE | ACTIVE | PROTECTED |
                         CYCLIC | SPORADIC
      class_parameters_declaration
      simple_ods    [-- with no real-time attributes ]
```

```
instance_ods ::=
   OBJECT object_name IS INSTANCE OF object_name
   [-- defined by a class_ods ]
      class_parameters_association
      simple_ods    [-- with all fields]
```

```
environment_ods ::=
   OBJECT object_name IS ENVIRONMENT
      object_description
      provided_interface
   END_OBJECT object_name
```

C.3 The Visible Part of the ODS

```
passive_visible_part ::=
   object_description
   passive_real_time_attributes | NONE
   implementation_constraints
   provided_interface
   required_interface
   data_flows
   exception_flows
```

```
active_visible_part ::=
   object_description
   active_real_time_attributes | NONE
   implementation_constraints
   provided_interface
   required_interface
   data_flows
   exception_flows
   active_obcs_specification
```

```
protected_visible_part ::=
   object_description
   protected_real_time_attributes | NONE
   implementation_constraints
   provided_interface
   required_interface
   data_flows
   exception_flows
   protected_obcs_specification

cyclic_visible_part ::=
   object_description
   cyclic_real_time_attributes | NONE
   implementation_constraints
   provided_interface
   required_interface
   data_flows
   exception_flows
   cyclic_obcs_specification

sporadic_visible_part ::=
   object_description
   sporadic_real_time_attributes | NONE
   implementation_constraints
   provided_interface
   required_interface
   data_flows
   exception_flows
   sporadic_obcs_specification
```

C.3.1 Real-Time Attributes

```
passive_real_time_attributes ::=
   REAL_TIME_ATTRIBUTES
      operation_RT_attributes
```

```
operation_RT_attributes ::=
  OPERATION operation_name [parameter_part][RETURN type_name] IS
    WCET mode_operation_execution_time_list
    BUDGET mode_operation_execution_time_list
  END_OPERATION operation_name
  {OPERATION operation_name [parameter_part][RETURN type_name] IS
    WCET mode_operation_execution_time_list
    BUDGET mode_operation_execution_time_list
  END_OPERATION operation_name }

active_real_time_attributes ::=
  REAL_TIME_ATTRIBUTES
    [VIRTUAL_NODE node_name ]
    operation_RT_attributes [-- for any ASER operations]


protected_real_time_attributes ::=
  REAL_TIME_ATTRIBUTES
    CEILING_PRIORITY mode_priority_list
    operation_RT_attributes

cyclic_real_time_attributes ::=
  REAL_TIME_ATTRIBUTES
    [VIRTUAL_NODE node_name ]
    PERIOD mode_period_list
    OFFSET mode_offset_list | NONE
    thread_real_time attributes
      operation_RT_attributes

sporadic_real_time_attributes ::=
  REAL_TIME_ATTRIBUTES
    [VIRTUAL_NODE node_name ]
    minimum_arrival_time | max_arrival_frequency |
      minimum_arrival_time max_arrival_frequency
    thread_real_time attributes
    operation_RT_attributes

minimum_arrival_time ::=
  MIN_ARRIVAL_TIME mode_arrival_time_list

max_arrival_frequency ::=
  MAX_ARRIVAL_FREQUENCY mode_arrival_frequency
```

```
thread_real_time_attributes ::=
   DEADLINE mode_deadline_list
   CEILING_PRIORITY mode_priority_list
   PRIORITY mode_priority_list
   PRECEDENCE_CONSTRAINTS
      BEFORE
         object_name_list | NONE
      AFTER
         object_name_list | NONE
   EXECUTION_TIME_TRANSFORMATION
      Ada_constant_integer | NONE
   IMPORTANCE mode_importance_list
   OPERATION thread IS
      WCET mode_thread_execution_time_list
      BUDGET mode_thread_execution_time_list
   END_OPERATION thread;

importance ::=
   HARD | SOFT | BACKGROUND
   [-- or some other appropriate definition ]
```

Mode information

```
mode_operation_execution_time_list ::=
   operation_name = mode_name =>
     Ada_constant_fine_duration { , mode_name =>
     Ada_constant_fine_duration}

mode_priority_list ::=
   mode_name => Ada_constant_integer {, mode_name =>
            Ada_constant_integer}

mode_period_list ::=
   mode_name => Ada_constant_fine_duration {, mode_name =>
            Ada_constant_fine_duration}

mode_offset_list ::=
   mode_name => Ada_constant_fine_duration {, mode_name =>
            Ada_constant_fine_duration}

mode_deadline_list ::=
   mode_name => Ada_constant_fine_duration {, mode_name =>
            Ada_constant_fine_duration}
```

mode_thread_execution_time_list ::=
  mode_name => Ada_constant_fine_duration {, mode_name =>
      Ada_constant_fine_duration}

mode_arrival_time_list ::=
  mode_name => Ada_constant_fine_duration {, mode_name =>
      Ada_constant_fine_duration}

mode_arrival_frequency ::=
  mode_name => Ada_constant_integer {, mode_name =>
      Ada_constant_integer}

mode_importance_list ::=
  mode_name => importance {, mode_name => importance}

C.3.2 Implementation constraints

implementation_constraints ::=
  IMPLEMENTATION_OR_SYNCHRONISATION_CONSTRAINTS
    Pragma_imp

Pragma_imp ::=
  [PRAGMA MAIN [operation_name]]
  [PRAGMA EXCEPTION Ada_procedure_declaration]

C.3.3 Provided Interface

provided_interface ::=
  PROVIDED_INTERFACE
    provided_types
    provided_constants
    provided_operation_sets
    provided_operations
    provided_exceptions

```
provided_types ::=
  TYPES
    type_declaration | NONE

type_declaration ::=
  Ada_identifier [Ada_discriminant_part] IS
    Ada_type_definition | [LIMITED] PRIVATE; informal_comment;
  {Ada_identifier [Ada_discriminant_part] IS
    Ada_type_definition | [LIMITED] PRIVATE; informal_comment;}

provided_constants ::=
  CONSTANTS
    constant_declaration | NONE

constant_declaration ::=
  Ada_constant_declaration ;
  { Ada_constant_declaration ; }

provided_operations ::=
  OPERATIONS
    operations | NONE

operations ::=
  operation_name [parameter_part][RETURN type_name] [MEMBER_OF
    operation_set_name];
  {operation_name [parameter_part][RETURN type_name] [MEMBER_OF
    operation_set_name];}

parameter_part ::=
  ( parameter_name : parameter_mode type_name [:= Ada_expression]
  {; parameter_name : parameter_mode type_name [:=
    Ada_expression] } )

parameter_name ::=
  name

parameter_mode ::=
  IN | IN OUT | OUT
```

provided_operation_sets ::=
    OPERATION_SETS
    operation_set_names | NONE

operation_set_names ::=

    operation_set_name [MEMBER_OF operation_set_name];
        {operation_set_name; [MEMBER_OF operation_set_name]}

provided_exceptions ::=
    EXCEPTIONS
    exception_list | NONE

exception_list ::=
    exception_name RAISED_BY operation_name {, operation_name};
    {exception_name RAISED_BY operation_name {, operation_name};}

C.3.4 Required Interface

required_interface ::=
    REQUIRED_INTERFACE
        [formal_parameters]
        required_objects

formal_parameters ::=
    FORMAL_PARAMETERS
        required_types
        required_constants
        required_operation_sets
        required_operations

```
required_objects ::=
  OBJECTS
    used_objects_list | NONE

used_objects_list ::=
  used_object
  {, used_object }

used_object ::=
  object_name
    required_types
    required_constants
    required_operation_sets
    required_operations
    required_exceptions


required_types ::=
  TYPES
    type_name_list | NONE

type_name_list ::=
    type_name {, type_name}

required_constants ::=
  CONSTANTS
    constant_name_list | NONE
constant_name_list ::=
    constant_name {, constant_name}

required_operation_sets ::=
  OPERATION_SETS
    operation_set_name {, operation_set_name)

required_operations ::=
  OPERATIONS
    used_operations | NONE
```

```
used_operations ::=
      operation_name [parameter_part][RETURN type_name];
   { operation_name [parameter_part][RETURN type_name]; }

required_exceptions ::=
   EXCEPTIONS
      used_exceptions | NONE

used_exceptions ::=
      exception_name  {, exception_name }
```

C.3.5 Dataflows

```
data_flows ::=
   DATAFLOWS
      dataflows | NONE

dataflows ::=
   data_name direction object_name;
   { data_name direction object_name; }

direction ::=
   => | <=> | <=
```

C.3.6 Exception Flows

```
exception_flows ::=
   EXCEPTION FLOWS
      exceptionflows | NONE

exceptionflows ::=
   exception_name <= object_name
   { exception_name <= object_name }
```

C.3.7 OBCS Specification

```
obcs_description ::=
   DESCRIPTION
      informal_text
```

```
active_obcs_specification ::=
  OBJECT_CONTROL_STRUCTURE
    obcs_description
    active_obcs_constrained_operations

active_obcs_constrained_operations ::=
  active_constrained_operations

active_constrained_operations ::=
  CONSTRAINED_OPERATIONS
    constrained_operations | NONE

constrained_operations ::=
    operation_name [constraint] ;
    { operation_name [constraint] ; }

constraint ::=
  CONSTRAINED_BY (constrained_label_text
          [functional_activation_constraint])

constrained_label_text ::=
  LSER | HSER | ASER | aser_interrupt | lser_toer | hser_toer

aser_interrupt ::=
  ASER_by_IT Ada_interrupt_declaration

lser_toer ::=
  LSER_TOER Ada_fine_duration_declaration

hser_toer ::=
  HSER_TOER Ada_fine_duration_declaration

functional_activation_constraint ::=
  informal_text

protected_obcs_specification ::=
  OBJECT_CONTROL_STRUCTURE
    obcs_description
    protected_obcs_constrained_operations

protected_obcs_constrained_operations ::=
  CONSTRAINED_OPERATIONS
    operation_name [protected_constraint] ;
    { operation_name [protected_constraint] ; }
```

protected_constraint ::=
  CONSTRAINED_BY (protected_constraint_labelled_text)

protected_constraint_labelled_text ::=
  PAER | PSER | pser_toer

pser_toer ::= PSER_TOER Ada_fine_duration_declaration
  functional_activation_constraint

cyclic_obcs_specification ::=
  OBJECT_CONTROL_STRUCTURE
    obcs_description
    cyclic_obcs_constrained_operations

cyclic_obcs_constrained_operations ::=
  CONSTRAINED_OPERATIONS
    cyclic_constrained_operations | NONE

cyclic_constrained_operations ::=
  operation_name cyclic_constraint;
  { operation_name cyclic_constraint; }

cyclic_constraint ::=
  CONSTRAINED_BY (cyclic_constraint_labelled_text)

cyclic_constraint_labelled_text ::=
  ASATC | LSATC | lsatc_toer | HSATC | toer_hsatc
  functional_activation_constraint

lsatc_toer ::=
  LSATC_TOER Ada_fine_duration_declaration

hsatc_toer ::=
  HSATC_TOER Ada_fine_duration_declaration

sporadic_obcs_specification ::=
  OBJECT_CONTROL_STRUCTURE
    obcs_description
    sporadic_obcs_constrained_operations

sporadic_obcs_constrained_operations ::=
  CONSTRAINED_OPERATIONS
    sporadic_constrained_operations | NONE

sporadic_constrained_operations ::=
operation_name cyclic_sporadic_constraint;
{ operation_name cyclic_sporadic_constraint; }

sporadic_constraint ::=
CONSTRAINED_BY (sporadic_constraint_labelled_text)

sporadic_constraint_labelled_text ::=
ASATC | LSATC | lsatc_toer | HSATC | hsatc_toer | ASER
functional_activation_constraint

C.4 The Hidden Part of the ODS

passive_hidden_part ::=
INTERNALS
terminal_passive_internals | non_terminal_passive_internal

active_hidden_part ::=
INTERNALS
terminal_active_internals | non_terminal_active_internal
active_obcs_implementation
opcs

protected_hidden_part ::=
INTERNALS
terminal_protected_internals | non_terminal_protected_internal
protected_obcs_implementation
opcs

cyclic_hidden_part ::=
INTERNALS
terminal_cyclic_internals | non_terminal_cyclic_internal
cyclic_obcs_implementation
opcs

sporadic_hidden_part ::=
INTERNALS
terminal_sporadic_internals | non_terminal_sporadic_internal
sporadic_obcs_implementation
opcs

C.4.1 Non Terminal Internals

```
non_terminal_passive_internals ::=
   non_terminal_declarations

non_terminal_active_internals ::=
   non_terminal_declarations
   non_terminal_active_obcs_implementation

non_terminal_protected_internals ::=
   non_terminal_declarations
   non_terminal_protected_obcs_implementation

non_terminal_cyclic_internals ::=::=
   non_terminal_declarations
   non_terminal_cyclic_obcs_implementation

non_terminal_sporadic_internals  ::=
   terminal_declarations
   non_terminal_sporadic_obcs_implementation

non_terminal_declarations ::=
   OBJECTS
      child_objects
   TYPES
      types_implementedby  | NONE
   DATA
      NONE
   CONSTANTS
      constants_implementedby | NONE
   OPERATIONS
      operations_implementedby | NONE
   EXCEPTIONS
      exception_implementedby | NONE

child_objects ::=
   object_name_list

types_implementedby ::=
   type_name IMPLEMENTED BY object_name.type_name;
   { type_name IMPLEMENTED BY object_name.type_name; }

constants_implementedby
   constant_name IMPLEMENTED BY object_name.constant_name;
   { constant_name IMPLEMENTED BY object_name.constant_name; }
```

```
operations_implementedby
   operation_name IMPLEMENTED BY object_name.operation_name;
   { operation_name IMPLEMENTED BY object_operation.type_name; }

exceptions_implementedby
   exception_name IMPLEMENTED BY object_name.exception_name;
   { exception_name IMPLEMENTED BY object_exception.type_name; }

non_terminal_active_obcs_implementation ::=
   ACTIVE_OBJECT_CONTROL_STRUCTURE
      IMPLEMENTED_BY object_name_list

non_terminal_protected_obcs_implementation ::=
   PROTECTED_OBJECT_CONTROL_STRUCTURE
      IMPLEMENTED_BY object_name_list

non_terminal_cyclic_obcs_implementation ::=
   CYCLIC_OBJECT_CONTROL_STRUCTURE
      IMPLEMENTED_BY object_name_list

non_terminal_sporadic_obcs_implementation ::=
   SPORADIC_OBJECT_CONTROL_STRUCTURE
      IMPLEMENTED_BY  object_name_list
```

C.4.2 Terminal Internals

```
terminal_passive_internals ::=
   terminal_declarations
   opcs

terminal_active_internals ::=
   terminal_declarations
   terminal_active_obcs_implementation
   opcs
```

terminal_protected_internals ::=
  terminal_declarations
  terminal_protected_obcs_implementation
  protected_opcs

terminal_cyclic_internals ::=::=
  terminal_terminal_declarations
  cyclic_obcs_implementation
  opcs

terminal_sporadic_internals ::=
  terminal_declarations
  terminal_sporadic_obcs_implementation
  opcs

terminal_declarations ::=
  OBJECTS
    NONE
  TYPES
    internal_type_declaration | NONE
  DATA
    internal_data_declaration | NONE
  CONSTANTS
    internal_constant_declaration | NONE
  OPERATIONS
    operations
  EXCEPTIONS
    exception_list | NONE

internal_type_declaration ::=
  Ada_identifier [Ada_discriminant_part] IS
    Ada_type_definition; informal_comment;
  { Ada_identifier [Ada_discriminant_part] IS
    Ada_type_definition; informal_comment; }

internal_data_declaration ::=
    data_name : type_name ;
    { data_name : type_name ; }

internal_constant_declaration ::=
    Ada_constant_declaration;
    { Ada_constant_declaration; }

```
terminal_active_obcs_implementation ::=
  ACTIVE_OBJECT_CONTROL_STRUCTURE
    obcs_code

terminal_protected_obcs_implementation ::=
  PROTECTED_OBJECT_CONTROL_STRUCTURE
    obcs_code

terminal_cyclic_obcs_implementation ::=
  CYCLIC_OBJECT_CONTROL_STRUCTURE
    obcs_code

terminal_sporadic_obcs_implementation ::=
  SPORADIC_OBJECT_CONTROL_STRUCTURE
    obcs_code

obcs_code ::=
  CODE
    pseudo_code

pseudo_code ::=
  informal_text
```

C.4.3 OPCS

```
opcs ::=
  OPERATION_CONTROL_STRUCTURE
    op_c_s

op_c_s ::=
  single_op_c_s { single_op_c_s }

single_op_c_s ::=
  OPERATION operation_name [parameter_part][RETURN type_name] IS
    [MEMBER_OF operation_set_name]
    opcs_description
    opcs_code
  END_OPERATION operation_name
```

```
opcs_description ::=
   DESCRIPTION
      informal_text
   USED OPERATIONS
        used_operations | NONE
   PROPAGATED_EXCEPTIONS
      opcs_exceptions | NONE
   HANDLED_EXCEPTIONS
      opcs_exceptions | NONE

opcs_exception ::=
   exception_name; { exception_name; }

opcs_code ::=
   CODE
      pseudo_code

protected_opcs ::=
   OPERATION_CONTROL_STRUCTURE
      protected_op_c_s

protected_op_c_s ::=
   protected_single_op_c_s { protected_single_op_c_s }

protected_single_op_c_s ::=
   OPERATION operation_name [parameter_part]
         [RETURN type_name] IS [PROTECTED]
      [MEMBER_OF operation_set_name]
      opcs_description
      opcs_code
   END_OPERATION operation_name
```

C.5 Parameters of Class objects

```
class_parameters_declaration::=
  FORMAL_PARAMETERS
    TYPE
      formal_class_types
    CONSTANTS
      formal_class_constants
    OPERATION_SETS
      formal_class_operation_sets
    OPERATION
      formal_class_operations

formal_class_types ::=
  NONE | type_name; { type_name;}

formal_class_constants ::=
  NONE | constant_name; { constant_name;}

formal_class_operation_sets ::= NONE | operation_set_names

formal_class_operations ::= NONE | operations


class_parameters_association::=
  FORMAL_PARAMETERS
    TYPE
      formal_class_types => actual_class_types
    CONSTANTS
      formal_class_constants => actual_class_constants
    OPERATION_SETS
      formal_class_operation_sets => actual_class_operation_sets
    OPERATION
      formal_class_operations => actual_class_operations

actual_class_types ::=
  NONE | type_name; { type_name;}

actual_class_constants ::=
  NONE | constant_name; { constant_name;}

actual_class_operation_sets ::= NONE | operation_set_names

actual_class_operations ::= NONE | operations
```

Appendix D:
Textual Formalism — the ODS Definition

HRT-HOOD is a structured design notation and, in common with most other structured notations, has a textual representation as well as the graphical notation. This section describes the textual representation of the HRT-HOOD notation.

The descriptions of objects in HRT-HOOD is produced using an *Object Definition Skeleton* (ODS) whose fields are updated as the design of an object is more and more refined. Each field in an ODS is described by a language HRT-HOOD PDL (based on HOOD PDL). The formal syntax of the language is given in BNF form in Appendix C.

Conventions used in the ODS descriptions are:

- keywords are marked in upper case
- comments to the ODS are marked with the "--" notation
- optional entries are enclosed by "[..]".

The structure of the ODS for the HRT-HOOD object types are given in the following sections.

D.1 PASSIVE Objects

OBJECT Object_name IS PASSIVE

DESCRIPTION
Informal description of the object.

REAL_TIME_ATTRIBUTES
[-- Description of the Objects real-time attributes. One field for each operation.]
INTEGRITY
[-- implementation-defined integrity level]
OPERATION Operation_Name IS
WCET
The worst case execution times for the operation in each mode.
BUDGET
The budget times allowed for the operation in each mode.

```
    END_OPERATION Operation_Name

  PROVIDED_INTERFACE
    [-- Formal description of the resources provided to other objects.]
    TYPES
      [-- type definitions provided to other objects — in target
         language syntax.]
      List of (Type_declarations with textual description)
    CONSTANTS
      [-- constants provided to other objects — in target language syntax.]
      List of (Constant_declarations with textual description)
    OPERATION_SETS
      List of (Operation_set_name with textual description)
    OPERATIONS
      [-- operation and parameter definitions — in target
         language syntax.]
      List of (Operation_name(list of (Parameter : Mode type)
           [RETURN type] with textual description)
    EXCEPTIONS
      [-- exceptions that could be transmitted to other objects]
      List of (Exception_name RAISED_BY (list of (Operation_name))
      with textual description)

  REQUIRED_INTERFACE
    [-- Formal description of the resources required from other objects]
    FORMAL_PARAMETERS  [ -- only if descendant of a CLASS ]
      TYPES
        List of (Type_name)
      CONSTANTS
        List of (Constant_name)
      OPERATION_SETS
        List of (Operation_set_name with textual description)
      OPERATIONS
        List of (Operation_name(list of (Parameter : Mode type)
             [RETURN type]))
    OBJECTS [ -- for all required objects ]
      Object_name
        TYPES
          List of (Type_name)
        CONSTANTS
          List of (Constant_name)
        OPERATION_SETS
          List of (Operation_set_name)
        OPERATIONS
          List of (Operation_name(list of (Parameter : Mode type)
```

```
                [RETURN type]))
          EXCEPTIONS
            List of (Exception_name)
    DATAFLOWS
      [-- Description of the data-flow name and direction
        represented in the diagram]
      List of (data_name direction object_name with textual description)
    EXCEPTION_FLOWS
      [-- Description of the exception flow name
        represented in the diagram]
      List of (exception_name <= Used_object_name)

    INTERNALS
      [-- Formal description of the Interface Implementation]
      OBJECTS
        NONE | [-- Only for terminal objects]
        List of (Child_object_name with textual description)
        [-- Only for non-terminal objects]
      TYPES
        List of (Type_names IMPLEMENTED_BY
          Child_object_name.Type_name)  [-- Only for non-terminal objects]
        List of (Internal_type_name) with textual description)
        [-- for terminal object only]
      DATA
        NONE |
        List of (Data_name : Type_name) with textual description)
        [-- for terminal object only]
      CONSTANTS
        List of (Constant_name IMPLEMENTED_BY
          Child_object_name.Constant_name)
        [-- Only for non-terminal objects]
        List of (Internal_constant_name) with textual description)
        [-- for terminal object only]
      OPERATION_SETS
        NONE |
        List of (Operation_set_names IMPLEMENTED_BY
          Child_object_name.Operation_set_name)
        [-- Only for non-terminal objects]
      OPERATIONS
        List of (Operation_names IMPLEMENTED_BY
          Child_object_name.Operation_name)
        [-- Only for non-terminal objects]
        List of (Internal_operation_name(list of (Parameter : Mode type)
          [RETURN type])
          with textual description) [-- for terminal object only]
```

```
    EXCEPTIONS
      List of (Exception_names IMPLEMENTED_BY
        Child_object_name.Exception_name)
      [-- Only for non-terminal objects]
      List of (Exception_name RAISED_BY (list of (Operation_name))
          with textual description) [-- Only for terminal objects]
    OPERATION_CONTROL_STRUCTURE [-- for terminal object only]
      [-- One operation description for each provided or internal operation]
      OPERATION Operation_name(list of (Parameter : Mode type) IS
            [RETURN type]) [MEMBER_OF set_name]
        DESCRIPTION
          Textual description of the operation
        USED_OPERATIONS [-- for terminal object only]
          List of (Operation_name(list of (Parameter : Mode type)
              [RETURN type]))
        PROPAGATED_EXCEPTIONS
          List of (Propagated_Exception_Name)
        HANDLED_EXCEPTIONS
          List of (Handled_Exception_Name)
        CODE
          Pseudo-code of the operation
      END_OPERATION Operation_name
  END_OBJECT Object_name
```

D.2 ACTIVE Objects

```
OBJECT Object_name IS ACTIVE

  DESCRIPTION
    Informal description of the object.

  REAL_TIME_ATTRIBUTES
    [-- Description of the Objects real-time attributes.
    VIRTUAL_NODE
      YES | NO

  IMPLEMENTATION_CONSTRAINTS
    [-- Description of any synchronisation, control sequence or implementation
    constraint which could apply to exported or internal operations ]

  PROVIDED_INTERFACE
    [-- Formal description of the resources provided to other objects.]
    TYPES
      [-- type definitions provided to other objects —
        in target language syntax.]
      List of (Type_declarations with textual description)
```

```
  CONSTANTS
    [-- constants provided to other objects — in target language syntax.]
    List of (Constant_declarations with textual description)
  OPERATION_SETS
    List of (Operation_set_name with textual description)
  OPERATIONS
    [-- operation and parameter definitions — in target language syntax.]
    List of (Operation_name(list of (Parameter : Mode type) [RETURN type]
    with textual description)
  EXCEPTIONS
    [-- exceptions that could be transmitted to other objects]
    List of (Exception_name RAISED_BY (list of (Operation_name)) with
    textual description)

REQUIRED_INTERFACE
  [-- Formal description of the resources required from other objects]
  FORMAL_PARAMETERS  [ -- only if descendant of a CLASS ]
    TYPES
      List of (Type_name)
    CONSTANTS
      List of (Constant_name)
    OPERATION_SETS
      List of (Operation_set_name with textual description)
    OPERATIONS
      List of (Operation_name(list of (Parameter : Mode type)
          [RETURN type]))
  OBJECTS [ -- for all required objects ]
    Object_name
      TYPES
        List of (Type_name)
      CONSTANTS
        List of (Constant_name)
      OPERATION_SETS
        List of (Operation_set_name)
      OPERATIONS
        List of (Operation_name(list of (Parameter : Mode type)
            [RETURN type]))
      EXCEPTIONS
        List of (Exception_name)
DATAFLOWS
  [-- Description of the data-flow name and direction
    represented in the diagram]
  List of (data_name direction object_name with textual description)
EXCEPTION_FLOWS
  [-- Description of the exception flow name represented in the diagram]
```

List of (exception_name <= Used_object_name)

ACTIVE_OBJECT_CONTROL_STRUCTURE
[-- Description of the synchronisation between the operations.]
DESCRIPTION
Description of the synchronisation between the operations.
CONSTRAINED_OPERATIONS
List of (Operation_Name) CONSTRAINED_BY
(ACTIVE_Constrained_label_text)

INTERNALS
[-- Formal description of the Interface Implementation]
OBJECTS
NONE | [-- Only for terminal objects]
List of (Child_object_name with textual description)
[-- Only for non-terminal objects]
TYPES
List of (Type_names IMPLEMENTED_BY
Child_object_name.Type_name) [-- Only for non-terminal objects]
List of (Internal_type_name) with textual description)
[-- for terminal object only]
DATA
List of (Data_name : Type_name) with textual description)
[-- for terminal object only]
CONSTANTS
List of (Constant_name IMPLEMENTED_BY
Child_object_name.Constant_name) [-- Only for non-terminal objects]
List of (Internal_constant_name) with textual description)
[-- for terminal object only]
OPERATION_SETS
List of (Operation_set_names IMPLEMENTED_BY
Child_object_name.Operation_set_name)
[-- Only for non-terminal objects]
OPERATIONS
List of (Operation_names IMPLEMENTED_BY
Child_object_name.Operation_name)
[-- Only for non-terminal objects]
List of (Internal_operation_name(list of (Parameter : Mode type)
[RETURN type])
with textual description) [-- for terminal object only]
EXCEPTIONS
List of (Exception_names IMPLEMENTED_BY
Child_object_name.Exception_name)
[-- Only for non-terminal objects]

```
    List of (Exception_name RAISED_BY (list of (Operation_name))
        with textual description) [-- Only for terminal objects]
  ACTIVE_OBJECT_CONTROL_STRUCTURE
    CODE [-- Only for terminal objects]
      Pseudo code limited to synchronisation
      between constrained operations
      [-- for non-terminal objects this reduces to:]
    IMPLEMENTED_BY List of (child_object_name)
      [-- for non-terminal object only]"

  OPERATION_CONTROL_STRUCTURE [-- for terminal object only]
    [-- One operation description for each provided or internal operation]
    OPERATION Operation_name(list of (Parameter : Mode type) IS
          [RETURN type]) [MEMBER_OF set_name]
      DESCRIPTION
        Textual description of the operation
      USED_OPERATIONS [-- for terminal object only]
        List of (Operation_name(list of (Parameter : Mode type)
            [RETURN type]))
      PROPAGATED_EXCEPTIONS
        List of (Propagated_Exception_Name)
      HANDLED_EXCEPTIONS
        List of (Handled_Exception_Name)
      CODE
        Pseudo-code of the operation
    END_OPERATION Operation_name
END_OBJECT Object_name
```

D.3 PROTECTED Objects

```
OBJECT Object_name IS PROTECTED

  DESCRIPTION
    Informal description of the object.

  REAL_TIME_ATTRIBUTES
    [-- Description of the Objects real-time attributes.]
    CEILING_PRIORITY
      [-- Ceiling priority of the object in each mode]
      List of priorities
    INTEGRITY
      [ -- implementation-defined integrity level ]
    OPERATION Operation_Name IS
      [ -- one for each operation ]
      WCET
        The worst case execution times for the operation in each mode.
```

BUDGET
The budget times allowed for the operation in each mode.
PRIORITY [-- if object accessed remotely]
[-- priority of the network communication, one for each mode]
List of priorities
END_OPERATION Operation_Name

PROVIDED_INTERFACE
[-- Formal description of the resources provided to other objects.]
TYPES
[-- type definitions provided to other objects —
in target language syntax.]
List of (Type_declarations with textual description)
CONSTANTS
[-- constants provided to other objects — in target language syntax.]
List of (Constant_declarations with textual description)
OPERATION_SETS
List of (Operation_set_name with textual description)
OPERATIONS
[-- operation and parameter definitions — in target language syntax.]
List of (Operation_name(list of (Parameter : Mode type) [RETURN type]
with textual description)
EXCEPTIONS
[-- exceptions that could be transmitted to other objects]
List of (Exception_name RAISED_BY (list of (Operation_name)) with
textual description)

REQUIRED_INTERFACE
[-- Formal description of the resources required from other objects]
FORMAL_PARAMETERS [-- only if descendant of a CLASS]
TYPES
List of (Type_name)
CONSTANTS
List of (Constant_name)
OPERATION_SETS
List of (Operation_set_name with textual description)
OPERATIONS
List of (Operation_name(list of (Parameter : Mode type)
[RETURN type]))
OBJECTS [-- for all required objects]
Object_name
TYPES
List of (Type_name)
CONSTANTS
List of (Constant_name)

OPERATION_SETS
List of (Operation_set_name)
OPERATIONS
List of (Operation_name(list of (Parameter : Mode type)
[RETURN type]))
EXCEPTIONS
List of (Exception_name)
DATAFLOWS
[-- Description of the data-flow name and direction
represented in the diagram]
List of (data_name direction object_name with textual description)
EXCEPTION_FLOWS
[-- Description of the exception flow name represented in the diagram]
List of (exception_name <= Used_object_name)
PROTECTED_OBJECT_CONTROL_STRUCTURE
[-- Description of the synchronisation between the operations.]
DESCRIPTION
Description of the synchronisation between the operations.
CONSTRAINED_OPERATIONS
List of (Operation_Name)
[-- constrained operations shall not have RETURN option]
CONSTRAINED_BY (PROTECTED_Constrained_label_text)

INTERNALS
[-- Formal description of the Interface Implementation]
OBJECTS
NONE | [-- Only for terminal objects]
List of (Child_object_name with textual description)
[-- Only for non-terminal objects]
TYPES
List of (Type_names IMPLEMENTED_BY
Child_object_name.Type_name) [-- Only for non-terminal objects]
List of (Internal_type_name) with textual description)
[-- for terminal object only]
DATA
List of (Data_name : Type_name) with textual description)
[-- for terminal object only]
CONSTANTS
List of (Constant_name IMPLEMENTED_BY
Child_object_name.Constant_name)
[-- Only for non-terminal objects]
List of (Internal_constant_name) with textual description)
[-- for terminal object only]
OPERATION_SETS
List of (Operation_set_names IMPLEMENTED_BY

```
      Child_object_name.Operation_set_name)
    [-- Only for non-terminal objects]
  OPERATIONS
    List of (Operation_names IMPLEMENTED_BY
      Child_object_name.Operation_name)
    [-- Only for non-terminal objects]
    List of (Internal_operation_name(list of (Parameter : Mode type)
      [RETURN type]) with textual description)
    [-- for terminal object only]
  EXCEPTIONS
    List of (Exception_names IMPLEMENTED_BY
      Child_object_name.Exception_name)
    [-- Only for non-terminal objects]
    List of (Exception_name RAISED_BY (list of (Operation_name))
        with textual description) [-- Only for terminal objects]
  PROTECTED_OBJECT_CONTROL_STRUCTURE
    [-- Description of the synchronisation between the operations.]
    CODE [-- Only for terminal objects]
      Pseudo code limited to synchronisation between constrained operations
      [-- for non-terminal objects this reduces to:]
    IMPLEMENTED_BY Child_object_name [-- Only for non-terminal objects]

  OPERATION_CONTROL_STRUCTURE [-- for terminal object only]
    [-- One operation description for each provided or internal operation]
    OPERATION Operation_name(list of (Parameter : Mode type) IS
        [RETURN type]) [PROTECTED]
      DESCRIPTION
        Textual description of the operation
      USED_OPERATIONS [-- for terminal object only]
        List of (Operation_name(list of (Parameter : Mode type)
          [RETURN type]))
      PROPAGATED_EXCEPTIONS
        List of (Propagated_Exception_Name)
      HANDLED_EXCEPTIONS
        List of (Handled_Exception_Name)
      CODE
        Pseudo-code of the operation
    END_OPERATION Operation_name
END_OBJECT Object_name
```

D.4 CYCLIC Objects

```
OBJECT Object_name IS CYCLIC

  DESCRIPTION
    Informal description of the object.
```

REAL_TIME_ATTRIBUTES

[-- Description of the Objects real-time attributes.]

PERIOD

[-- Description of the periodicity of the object, one for each mode]

List of period times

OFFSET

[-- Offset before the object can start executing its cyclic operations, one for each mode]

List of offset times

DEADLINE

[-- Description of the deadline of the object, one for each mode]

List of deadline times

PRIORITY

[-- Priority of the object, one for each mode]

List of priorities

CEILING_PRIORITY

[-- Ceiling priority of the object in each mode]

List of priorities

PRECEDENCE CONSTRAINTS

[-- Description of any precedence constraints]

BEFORE list of objects

AFTER list of objects

EXECUTION TIME TRANSFORMATION

[-- Number of sections of code that the THREAD should be transformed, one for each mode] list of integers

IMPORTANCE

[-- Importance of the object to the system]

list of HARD | SOFT | BACKGROUND

INTEGRITY

[-- implementation-defined integrity level]

OPERATION Operation_Name IS

[-- One field for each operation including the THREAD operation called each period.]

WCET

The worst case execution times for the operation in each mode.

BUDGET

The budget times allowed for the operation in each mode.

END_OPERATION Operation_Name

PROVIDED_INTERFACE

[-- Formal description of the resources provided to other objects.]

TYPES

[-- type definitions provided to other objects — in target language syntax.]

List of (Type_declarations with textual description)

CONSTANTS
[-- constants provided to other objects — in target language syntax.]
List of (Constant_declarations with textual description)
OPERATION_SETS
List of (Operation_set_name with textual description)
OPERATIONS
[-- operation and parameter definitions — in target language syntax.]
List of (Operation_name(list of (Parameter : Mode type) [RETURN type]
with textual description)
EXCEPTIONS
[-- exceptions that could be transmitted to other objects]
List of (Exception_name RAISED_BY (list of (Operation_name)) with
textual description)

REQUIRED_INTERFACE
[-- Formal description of the resources required from other objects]
FORMAL_PARAMETERS [-- only if descendant of a CLASS]
TYPES
List of (Type_name)
CONSTANTS
List of (Constant_name)
OPERATION_SETS
List of (Operation_set_name with textual description)
OPERATIONS
List of (Operation_name(list of (Parameter : Mode type)
[RETURN type]))
OBJECTS [-- for all required objects]
Object_name
TYPES
List of (Type_name)
CONSTANTS
List of (Constant_name)
OPERATION_SETS
List of (Operation_set_name)
OPERATIONS
List of (Operation_name(list of (Parameter : Mode type)
[RETURN type]))
EXCEPTIONS
List of (Exception_name)
DATAFLOWS
[-- Description of the data-flow name and direction
represented in the diagram]
List of (data_name direction object_name with textual description)
EXCEPTION_FLOWS
[-- Description of the exception flow name represented in the diagram]

List of (exception_name <= Used_object_name)

CYCLIC_OBJECT_CONTROL_STRUCTURE
[-- Description of the synchronisation between the operations.]
DESCRIPTION
Description of the synchronisation between the operations.
CONSTRAINED_OPERATIONS
List of (Operation_Name)
[-- constrained operations shall not have RETURN option]
CONSTRAINED_BY (CYCLIC_Constrained_label_text)

INTERNALS
[-- Formal description of the Interface Implementation]
OBJECTS
NONE | [-- Only for terminal objects]
List of (Child_object_name with textual description)
[-- Only for non-terminal objects]
TYPES
List of (Type_names IMPLEMENTED_BY
Child_object_name.Type_name) [-- Only for non-terminal objects]
List of (Internal_type_name) with textual description)
[-- for terminal object only]
DATA
List of (Data_name : Type_name) with textual description)
[-- for terminal object only]
CONSTANTS
List of (Constant_name IMPLEMENTED_BY
Child_object_name.Constant_name)
[-- Only for non-terminal objects]
List of (Internal_constant_name) with textual description)
[-- for terminal object only]
OPERATION_SETS
List of (Operation_set_names IMPLEMENTED_BY
Child_object_name.Operation_set_name)
[-- Only for non-terminal objects]
OPERATIONS
List of (Operation_names IMPLEMENTED_BY
Child_object_name.Operation_name)
[-- Only for non-terminal objects]
List of (Internal_operation_name(list of (Parameter : Mode type)
[RETURN type])
with textual description) [-- for terminal object only]
EXCEPTIONS
List of (Exception_names IMPLEMENTED_BY
Child_object_name.Exception_name)

```
        [-- Only for non-terminal objects]
        List of (Exception_name RAISED_BY (list of (Operation_name))
                with textual description) [-- Only for terminal objects]
    CYCLIC_OBJECT_CONTROL_STRUCTURE
        [-- Description of the synchronisation between the operations.]
        CODE [-- Only for terminal objects]
            Pseudo code limited to synchronisation between
            constrained operations
            [-- for non-terminal objects this reduces to:]
        IMPLEMENTED_BY
            Child_object_name
        OPERATION_CONTROL_STRUCTURE [-- for terminal object only]
            [-- One operation description for each provided or internal operation]

        OPERATION Operation_name(list of (Parameter : Mode type) IS
                [RETURN type]) [MEMBER_OF set_name]
            DESCRIPTION
                Textual description of the operation
            USED_OPERATIONS [-- for terminal object only]
                List of (Operation_name(list of (Parameter : Mode type)
                    [RETURN type]))
            PROPAGATED_EXCEPTIONS
                List of (Propagated_Exception_Name)
            HANDLED_EXCEPTIONS
                List of (Handled_Exception_Name)
            CODE [-- for terminal objects only]
                Pseudo-code of the operation
        END_OPERATION Operation_name

        OPERATION THREAD IS
            [-- the code executed each cycle,
            terminal objects only]
            DESCRIPTION
                Description of the code executed each period
            USED_OPERATIONS [-- for terminal object only]
                List of (Operation_name(list of (Parameter : Mode type)
                    [RETURN type]))
            CODE
                Pseudo-code of the operation
        END_OPERATION THREAD
END_OBJECT Object_name
```

D.5 SPORADIC Objects

OBJECT Object_name IS SPORADIC

DESCRIPTION
Informal description of the object.

REAL_TIME_ATTRIBUTES
[-- Description of the Objects real-time attributes.]
MIN_ARRIVAL_TIME AND/OR MAX_ARRIVAL_FREQUENCY
[-- Description of the minimum arrival or maximum frequency of the sporadic, one per mode]
List of arrival times or arrival frequencies
OFFSET
[-- Offset before the object can start executing its sporadic operations, one for each mode]
List of offset times
DEADLINE
[-- Description of the deadline of the object, one per mode]
List of deadline times
PRIORITY
[-- Priority of the object, one per mode]
List of priorities.
CEILING_PRIORITY
[-- Ceiling priority of the object in each mode]
List of priorities
PRECEDENCE CONSTRAINTS
[-- Description of any precedence constraints]
BEFORE list of objects
AFTER list of objects
EXECUTION TIME TRANSFORMATION
[-- Number of sections of code that the THREAD should be transformed, one for each mode] list of integers
IMPORTANCE
[-- Importance of the object to the system, one per mode]
list of HARD | SOFT | BACKGROUND
INTEGRITY
[-- implementation-defined integrity level]
OPERATION Operation_Name IS
[-- One field for each operation including the internal operation called each activation.]
WCET
The worst case execution times for the operation in each mode.
BUDGET
The budget times allowed for the operation in each mode.
PRIORITY [-- if object accessed remotely]

```
      [-- priority of the network communication,  one for each mode]
      List of priorities
  END_OPERATION Operation_Name

PROVIDED_INTERFACE
  [-- Formal description of the resources provided to other objects.]
  TYPES
    [-- type definitions provided to other objects —
      in target language syntax.]
    List of (Type_declarations with textual description)
  CONSTANTS
    [-- constants provided to other objects — in target language syntax.]
    List of (Constant_declarations with textual description)
  OPERATION_SETS
    List of (Operation_set_name with textual description)
  OPERATIONS
    [-- operation and parameter definitions — in target language syntax.]
    List of (Operation_name(list of (Parameter : Mode type) [RETURN type]
    with textual description)
  EXCEPTIONS
    [-- exceptions that could be transmitted to other objects]
    List of (Exception_name RAISED_BY (list of (Operation_name)) with
    textual description)

REQUIRED_INTERFACE
  [-- Formal description of the resources required from other objects]
  FORMAL_PARAMETERS   [ -- only if descendant of a CLASS ]
    TYPES
      List of (Type_name)
    CONSTANTS
      List of (Constant_name)
    OPERATION_SETS
      List of (Operation_set_name with textual description)
    OPERATIONS
      List of (Operation_name(list of (Parameter : Mode type)
        [RETURN type]))
  OBJECTS [ -- for all required objects ]
    Object_name
      TYPES
        List of (Type_name)
      CONSTANTS
        List of (Constant_name)
      OPERATION_SETS
        List of (Operation_set_name)
      OPERATIONS
```

List of (Operation_name(list of (Parameter : Mode type)
[RETURN type]))
EXCEPTIONS
List of (Exception_name)
DATAFLOWS
[-- Description of the data-flow name and direction
represented in the diagram]
List of (data_name direction object_name with textual description)
EXCEPTION_FLOWS
[-- Description of the exception flow name represented in the diagram]
List of (exception_name <= Used_object_name)

SPORADIC_OBJECT_CONTROL_STRUCTURE
[-- Description of the synchronisation between the operations.]
DESCRIPTION
Description of the synchronisation between the operations.
CONSTRAINED_OPERATIONS
List of (Operation_Name)
[-- constrained operations shall not have RETURN option]
CONSTRAINED_BY (SPORADIC_Constrained_label_text)

INTERNALS
[-- Formal description of the Interface Implementation]
OBJECTS
NONE | [-- Only for terminal objects]
List of (Child_object_name with textual description)
[-- Only for non-terminal objects]
TYPES
List of (Type_names IMPLEMENTED_BY
Child_object_name.Type_name) [-- Only for non-terminal objects]
List of (Internal_type_name) with textual description)
[-- for terminal object only]
DATA
List of (Data_name : Type_name) with textual description)
[-- for terminal object only]
CONSTANTS
List of (Constant_name IMPLEMENTED_BY
Child_object_name.Constant_name) [-- Only for non-terminal objects]
List of (Internal_constant_name) with textual description)
[-- for terminal object only]
OPERATION_SETS
List of (Operation_set_names IMPLEMENTED_BY
Child_object_name.Operation_set_name)
[-- Only for non-terminal objects]
OPERATIONS

```
   List of (Operation_names IMPLEMENTED_BY
      Child_object_name.Operation_name)
   [-- Only for non-terminal objects]
   List of (Internal_operation_name(list of (Parameter : Mode type)
      [RETURN type]) with textual description)
   [-- for terminal object only]
EXCEPTIONS
   List of (Exception_names IMPLEMENTED_BY
      Child_object_name.Exception_name)
   [-- Only for non-terminal objects]
   List of (Exception_name RAISED_BY (list of (Operation_name))
            with textual description) [-- Only for terminal objects]
SPORADIC_OBJECT_CONTROL_STRUCTURE
   [-- Description of the synchronisation between the operations.]
   CODE [-- Only for terminal objects]
      Pseudo code limited to synchronisation
      between constrained operations
      [-- for non-terminal objects this reduces to:]
   IMPLEMENTED_BY
      Child_object_name
   OPERATION_CONTROL_STRUCTURE [-- for terminal object only]
      [-- One operation description for each provided or internal operation]

   OPERATION Operation_name(list of (Parameter : Mode type) IS
         [RETURN type]) [MEMBER_OF set_name]
      DESCRIPTION
         Textual description of the operation
      USED_OPERATIONS [-- for terminal object only]
         List of (Operation_name(list of (Parameter : Mode type)
            [RETURN type]))
      PROPAGATED_EXCEPTIONS
         List of (Propagated_Exception_Name)
      HANDLED_EXCEPTIONS
         List of (Handled_Exception_Name)
      CODE [-- for terminal objects only]
         Pseudo-code of the operation
   END_OPERATION Operation_name

   OPERATION THREAD IS
      [-- the code executed each cycle,
      terminal objects only]
      DESCRIPTION
         Description of the code executed each period
      USED_OPERATIONS [-- for terminal object only]
         List of (Operation_name(list of (Parameter : Mode type)
```

```
              [RETURN type]))
          CODE
             Pseudo-code of the operation
       END_OPERATION THREAD
END_OBJECT Object_name
```

D.6 ENVIRONMENT Objects

```
OBJECT Object_name IS ENVIRONMENT

   DESCRIPTION
      Informal description of the object.
   PROVIDED_INTERFACE
      [-- Formal description of the resources provided to other objects.]
      TYPES
         [-- type definitions provided to other objects —
            in target language syntax.]
         List of (Type_declarations with textual description)
      CONSTANTS
         [-- constants provided to other objects — in target language syntax.]
         List of (Constant_declarations with textual description)
      OPERATION_SETS
         List of (Operation_set_name with textual description)
      OPERATIONS
         [-- operation and parameter definitions — in target language syntax.]
         List of (Operation_name(list of (Parameter : Mode type) [RETURN type]
         with textual description)
      EXCEPTIONS
         [-- exceptions that could be transmitted to other objects]
         List of (Exception_name RAISED_BY (list of (Operation_name)) with
         textual description)
END_OBJECT Object_name
```

D.7 CLASS Objects

```
   OBJECT Class_name IS
      CLASS PASSIVE|ACTIVE|CYCLIC|SPORADIC|PROTECTED
     PARAMETERS
        TYPES
           List of (Formal_Type_declarations with textual description)
        CONSTANTS
           List of (Formal_Constant_declarations with textual description)
        OPERATION_SETS
           List of (Formal_Operation_set_name with textual description)
        OPERATIONS
           List of (Formal_Operation_name(list of (Parameter : Mode type)
```

[RETURN type] with textual description)

[-- standard ODS for object type without the real-time attributes]
END_OBJECT Class_name

D.8 Instances of CLASS Objects

```
OBJECT Object_name IS INSTANCE OF Class_name
  PARAMETERS
  TYPES
    List of (Formal_Type_declarations =>
      Actual_Type_declarations with textual description)
  CONSTANTS
    List of (Formal_Constant_declarations =>
      Actual_Constant_declarations with textual description)
  OPERATION_SETS
    List of (Formal_Operation_set_name =>
      Actual_Operation_set_name with textual description)
  OPERATIONS
    List of (Formal_Operation_name =>
      Actual_Formal_Operation_name (list of (Parameter : Mode type)
      [RETURN type] with textual description)

    [ standard fields for the appropriate object type ]
```

Appendix E:
Device Driving

This appendix contains information used in the Mine Control Case Study (see Chapter 9).

Each external device is controlled by one or two registers: a control and status register(CSR) and a data buffer register(DBR). The following table (Figure E.1) indicates the hardware address of the interrupt location (if used) and the registers, of each device.

Device	Interrupt Address	CSR	DBR
High Water Sensor	16#40#	16#AA10#	-
Low Water Sensor	16#44#	16#AA12#	-
Water Flow Sensor	-	16#AA14#	-
Pump Motor	-	16#AA16#	-
CH4 Sensor	-	16#AA18#	16#AA1A#
CO Sensor	-	16#AA1C#	16#AA1E#
Air Flow Sensor	-	16#AA20#	-

Figure E.1: Hardware Device Registers

Each register is 16 bits long. The control and status registers have the format shown in Figure E.2. For the CH4 and CO ADC devices the "device operation" bit is used to start the conversion (only one input channel is assumed).

15		10		7	6		0 bit
Device Error		Device Operation		Done	Interrupt Enable		Device Enable

Figure E.2: Control and Status Register Bits

The two data buffer registers return a scaled integer value which represents the level of gas in the surrounding atmosphere. The ADC provides a 10-bit reading which is mapped on to the least significant bits of the register. A reading between 0 and 1023 is therefore possible. For the methane, a value of 400 (on this scale) is considered too high. The associated value for carbon monoxide is 600.

OBJECT device_register IS PASSIVE

DESCRIPTION

This object simply defines some standard types associated with a typical device register.

REAL_TIME_ATTRIBUTES

NONE -- types only

PROVIDED_INTERFACE

CONSTANTS

```
word : constant := 2;       -- two bytes in a word
one_word : constant := 16; -- 16 bits in a word
```

TYPES

```
device_error is (clear, set);
device_operation is (clear, set);
interrupt_status is (i_disabled, i_enabled);
device_status is (d_disabled, d_enabled);

-- register type itself
csr is
    record
      error_bit  : device_error;
      operation  : device_operation;
      done       : boolean;
      interrupt  : interrupt_status;
      device     : device_status;
    end record;

-- bit representation of the register field
for device_error use (clear => 0, set => 1);
```

```
        for device_operation use (clear => 0, set => 1);
        for interrupt_status use (i_disabled => 0,
                                  i_enabled => 1);
        for device_status use (d_disabled => 0,
                               d_enabled => 1);
        for csr use
            record at mod word;
              error_bit  at 0 range 15 ..15;
              operation  at 0 range 10 ..10;
              done       at 0 range 7 .. 7;
              interrupt  at 0 range 6 .. 6;
              device     at 0 range 0 .. 0;
            end record;
        for csr'size use one_word;
      OPERATIONS
        NONE
      OPERATION_SETS
        NONE
    EXCEPTIONS
      NONE

  REQUIRED_INTERFACE
    OBJECTS
      NONE

  DATAFLOW
    NONE
  EXCEPTION_FLOWS
    NONE

  INTERNALS
    OBJECTS
      NONE
    TYPES
      NONE
    DATA
      NONE
    OPERATIONS
      NONE
  OPERATION_CONTROL_STRUCTURE
    NONE
END_OBJECT device_register
```

The resulting generated package is:

```
package device_register_types is

  word : constant := 2;       -- two bytes in a word
  one_word : constant := 16; -- 16 bits in a word
  -- register field types
  type device_error is (clear, set);
  type device_operation is (clear, set);
  type interrupt_status is (i_disabled, i_enabled);
  type device_status is (d_disabled, d_enabled);

  -- register type itself
  type csr is
    record
      error_bit  : device_error;
      operation  : device_operation;
      done       : boolean;
      interrupt  : interrupt_status;
      device     : device_status;
    end record;
  -- bit representation of the register field
  for device_error use (clear => 0, set => 1);
  for device_operation use (clear => 0, set => 1);
  for interrupt_status use (i_disabled => 0,
                            i_enabled => 1);
  for device_status use (d_disabled => 0,
                         d_enabled => 1);
  for csr use
    record at mod word;
      error_bit  at 0 range 15 ..15;
      operation  at 0 range 10 ..10;
      done       at 0 range 7 .. 7;
      interrupt  at 0 range 6 .. 6;
      device     at 0 range 0 .. 0;
    end record;
  for csr'size use one_word;
end device_register_types;
```

References

1. *York Ada Compiler Environment (York ACE) Reference Guide,* York Software Engineering Limited, 1991. (Release 5.1)
2. European Space Agency, ''HOOD User Manual Issue 3.0,'' WME/89-353/JB, December 1989.
3. European Space Agency, ''HOOD Reference Manual Issue 3.0,'' WME/89-173/JB, September 1989.
4. European Space Agency, ''HOOD Reference Manual Issue 3.1,'' HRM/91-07/V3.1, July 1991.
5. C. Atkinson, T. Moreton, and A. Natali, *Ada for Distributed Systems,* Ada Companion Series, Cambridge University Press, 1988.
6. N. Audsley, A. Burns, M. Richardson, K. Tindell, and A. Wellings, ''Applying New Scheduling Theory to Static Priority Pre-emptive Scheduling,'' *Software Engineering Journal*, vol. 8, no. 5, pp. 284-292, September 1993.
7. N.C. Audsley, A. Burns, and A.J. Wellings, ''Deadline Monotonic Scheduling Theory and Application,'' *Control Engineering Practice*, vol. 1, no. 1, pp. 71-78, 1993.
8. N.C. Audsley, A. Burns, M.F. Richardson, and A.J. Wellings, ''STRESS: A Simulator for Hard Real-Time Systems,'' *Software-Practice and Experience*, vol. 24, no. 6, pp. 543-564, June 1994.
9. C. Bailey, ''Survey of Typical Space Applications,'' Task 6 Deliverable on ESTEC Contract 9198/90/NL/SF, British Aerospace Space Systems Ltd., September 1991.
10. C. Bailey, ''Software Requirements Document for the Olympus AOCS,'' Task 10 Deliverable on ESTEC Contract 9198/90/NL/SF, British Aerospace Space Systems Ltd., March 1992.
11. C.M. Bailey, A. Burns, E. Fyfe, F. Gomez-Molinero, and A.J. Wellings, ''Implementing Hard Real-time Systems: A Case Study,'' *Proceeding*

International Symposium on Real-time Embedded Processing for Space Applications, Les Saintes-Maries-de-la-Mer, France, November 1992.

12. C.M. Bailey, E. Fyfe, T Vardanega, and A.J. Wellings, "The Use of Preemptive Priority-Based Scheduling in Space Applications," *Proceedings Real Time Systems Symposium, IEEE Computer Society*, pp. 253-257, North Carolina, December 1993.

13. J A. Bannister and K.S. Trivedi, "Task Allocation in Fault-Tolerant Distributed Systems," *Acta Informatica*, vol. 20, pp. 261-281, 1983.

14. P.A. Barrett, A.M. Hilborne, P. Verissimo, L. Rodrigues, P.G. Bond, D.T. Seaton, and N.A. Speirs, "The Delta-4 Extra Performance Architecture(XPA)," *Digest of Papers, The Twentieth Annual International Symposium on Fault-Tolerant Computing*, pp. 481-488, Newcastle, June 1990.

15. A.D. Birrell and B.J. Nelson, "Implementing Remote Procedure Calls," *ACM Transactions on Computer Systems*, vol. 2, no. 1, pp. 39-59, 1984.

16. P. Brinch-Hansen, *Operating System Principles*, Prentice-Hall, New Jersey, July 1973.

17. R.J.A. Buhr, *System Design with Ada*, Prentice-Hall International, 1984.

18. A. Burns and A. M. Lister, "An Architectural Framework for Timely and Reliable Distributed Information Systems(TARDIS): Description and Case Study," YCS.140, Department of Computer Science, University of York, 1990.

19. A. Burns and A.J. Wellings, *Real-time Systems and their Programming Languages*, Addison Wesley , 1990.

20. A. Burns and A.J. Wellings, "Real-time Ada: Outstanding Problem Areas," *Proceedings of the 3nd International Workshop on Real Time Ada Issues, ACM Ada Letters, Ada Letters*, vol. X, no. 4, pp. 5-14, 1990.

21. A. Burns and T.J. Quiggle, "Effective Use of Abort in Programming Mode Changes," *Ada Letters*, 1990.

22. A. Burns and A.J. Wellings, "Usability of the Ada Tasking Model," *Proceedings of the 3nd International Workshop on Real Time Ada Issues, ACM Ada Letters, Ada Letters*, vol. X, no. 4, pp. 49-56, 1990.

23. A. Burns, "Scheduling Hard Real-Time Systems: A Review," *Software Engineering Journal*, vol. 6, no. 3, pp. 116-128, 1991.

24. A. Burns and A.J. Wellings, "Development of a Design Methodology," Task 3 Deliverable on ESTEC Contract 9198/90/NL/SF, Department of Computer Science, University of York, September 1991.

25. A. Burns and A. M. Lister, "A Framework for Building Dependable Systems," *Computer Journal*, vol. 34, no. 2, pp. 173-181, 1991.

26. A. Burns and A.J. Wellings, "Definition of Tools," Task 4 Deliverable on ESTEC Contract 9198/90/NL/SF, Department of Computer Science, University of York, September 1991.

27. A. Burns, A.J. Wellings, C.M. Bailey, and E. Fyfe, "The Olympus Attitude and Orbital Control System: A Case Study in Hard Real-time System Design and Implementation," in *Ada sans frontieres Proceedings of the 12th Ada-Europe Conference, Lecture Notes in Computer Science 688*, pp. 19-35, Springer-Verlag, 1993.

28. A. Burns, A.J. Wellings, C.M. Bailey, and E. Fyfe, "The Olympus Attitude and Orbital Control System: A Case Study in Hard Real-time System Design and Implementation," YCS 190, Department of Computer Science, University of York, 1993.

29. A. Burns and A.J. Wellings, "Measuring, Monitoring and Enforcing CPU Execution Time Usage ," *Proceedings of the 6th International Workshop on Real Time Ada Issues, Ada Letters*, vol. XIII, no. 2, pp. 54-64, March/April 1993. (Also appears in Ada User 13(2) pp 73-78)

30. A. Burns, A.J. Wellings, and A.D. Hutcheon, "The Impact of an Ada Run-time System's Performance Characteristics on Scheduling Models," in *Ada sans frontieres Proceedings of the 12th Ada-Europe Conference, Lecture Notes in Computer Science 688*, pp. 240-248, Springer-Verlag, 1993.

31. A. Burns and A.J. Wellings, "Bridging the Real-time Gap between Ada 83 and Ada 9X," in *Ada Year Book*, ed. C. Loftus, pp. 71-86, IOS Press, 1993.

32. A. Burns and A.J. Wellings, K. Tindell, A. Burns, and A.J. Wellings, "Allocating Real-Time Tasks (An NP-Hard Problem made Easy)," *Real-Time Systems*, vol. 4, no. 2, pp. 145-165, June 1992. Also appears as YCS 199, Department of Computer Science, University of York

33. A. Burns and A. J. Wellings, "Implementing Analysable Hard Real-Time Sporadic Tasks in Ada 9X," *Ada Letters*, vol. 14, no. 1, pp. 38-49, January 1994.

34. J.P. Calvez, *Embedded Real-time Systems: A Specification and Design Methodology*, Wiley, 1993.

35. G. Chen and J. Yur, "A Branch-and-Bound-with-Underestimates Algorithm for the Task Assignment Problem with Precedence Constraint," *10th International Conference on Distributed Computing Systems*, pp. 494-501, 1990.

36. G.W. Cherry, *Pamela Designers Handbook,* Thought Tools Incorporated, 1986.

37. J.E. Cooling, *Software Design for Real-time Systems,* Chapman and Hall, 1991.

38. A. Damm, J. Reisinger, W. Schwabl, and H. Kopetz, "The Real-Time Operating System of MARS," *ACM Operating Systems Review*, vol. 23, no. 3 (Special Issue), pp. 141-157, 1989.

39. S. Davari and S.K. Dhall, "An On-line Algorithm for Real-Time Task Allocation," in *Proceedings 7th IEEE Real-Time Systems Symposium*, pp. 194-200, December 1986.

40. Minstry of Defence, "Hazard Analysis and Safety Classification of the Computer and Programmable Electronic System Elements of Defence Equipment," Interim Defence Standard, 00-56/Issue 1, April 1991.

41. Hull, M.E.C., O'Donoghue, P.G., and Hagan, B.J., "Development methods for real-time systems," *Computer Journal*, vol. 34, no. 2, pp. 164-72, April 1991. Comput. J. (UK)

42. J. E. Dobson and J. A. McDermid, "An Investigation into Modelling and Categorisation of Non-Functional Requirements," YCS.141, Department of Computer Science, University of York, 1990.

43. C.H. Forsyth, "Implementation of the Worst-Case Execution Time Analyser," Task 8 Volume E, Deliverable on ESTEC Contract 9198/90/NL/SF, York Software Engineering Limited, University of York, June 1992.

44. F. Gomez-Molinero, "Software Development Guidelines for the OBDH Environment, A Tutorial Example of a Generic On-Board Software Architecture," Work Package 2 Report, ESTEC Study Contract no.124386/92/NL/WK, TEICE Control report C-2422/0002, June 1993.

45. J.B. Goodenough and L. Sha, "The Priority Ceiling Protocol," Proceedings of the 2nd International Workshop on Real Time Ada Issues, Devon, 1988.

46. Ada Run Time Environment Working Group, "A Catalog of Interface Features and Options for the Ada Run Time Environment Release 3.0," *Ada Letters Special Issue*, vol. XI, no. 8, 1991.

47. Klein, M. H., Ralya, T. A., Pollak, B., Obenza, R., and Harbour, M. G., *A Practitioner's Handbook for Real-Time Analysis: A Guide to Rate Monotonic A nalysis for Real-Time Systems*, Kluwer Academic Publishers , 1993.

48. C.A.R. Hoare, "Monitors - An Operating System Structuring Concept," *CACM*, vol. 17, no. 10, pp. 549-557, October 1974.

49. A.D. Hutcheon and A.J. Wellings, "The Virtual Node Approach to Designing Distributed Ada Programs," *Ada User*, vol. 9, no. Supplement, pp. 35-42, December 1988.

50. A.D. Hutcheon and A.J. Wellings, "The York Distributed Ada Project," in *Distributed Ada: developments and experiences*, ed. J. Bishop, pp. 67-104, Cambridge University Press, 1990.

51. Intermetrics, "Draft Ada 9X Mapping Document, Volume II, Mapping Specification," Ada 9X Project Report, August 1991.

52. K. Jackson, "Mascot 3 and Ada," *Software Engineering Journal*, vol. 1, no. 3, pp. 121-135, May 1986.

53. M.A. Jackson, *Principles of Program Design*, Academic Press Inc, 1975.

54. Kopetz, H., Zainlinger, R., Fohler, G., Kantz, H., Puschner, P., and Schutz, W., "The design of real-time systems: from specification to implementation and verification," *Software Engineering Journal*, vol. 6, no. 3, pp. 72-82, May 1991. Softw. Eng. J. (UK)

55. H. Kopetz, "Design Principles for Fault Tolerant Real Time Systems," MARS Report, Institut flir Technische Informatik, 8/85/2, 1985 .

56. J. Kramer, J. Magee, M.S. Sloman, and A.M. Lister, "CONIC: an Integrated Approach to Distributed Computer Control Systems," *IEE Proceedings (Part E)*, vol. 180, no. 1, pp. 1-10, Jan 1983.

57. J. Kramer and J. Magee, "A Model for Change Management," *Proceedings of the IEEE Distributed Computing Systems in the '90s Conference*, September 1988.

58. B.W. Lampson, "Remote Procedure Calls," in *Lecture Notes in Computer Science, Vol. 105*, pp. 365-370, Springer-Verlag, 1981.

59. J.C. Laprie, "Dependability: A Unified Concept for Reliable Computing and Fault Tolerance," in *Resilient Computer Systems*, ed. T. Anderson, pp. 1-28, Collins and Wiley, 1989.

60. R.J. Lauber(Ed), "EPOS Overview: Short Account of the Main Featurs," GPP Gesellschaft fur ProzeBrechnerprogrammierung mbH, Kolpingring 18a, D-8024 Oberhaching bei M unchen, January 1990.

61. R J Lauber, "Forecasting Real-time Behaviour During Software Design Using a CASE Environment," *The Journal of Real-Time Systems*, vol. 1, pp. 61-76, 1989.

62. J.P. Lehoczky, L. Sha, and J.K. Strosnider, *Enhancing Aperiodic Responsiveness in Hard Real-Time Environment*, San Jose, California, December 1987. Proc. 8th IEEE Real-Time Systems Symposium

63. J.Y.T. Leung and J. Whitehead, "On the Complexity of Fixed-Priority Scheduling of Periodic, Real-Time Tasks," *Performance Evaluation (Netherlands)*, vol. 2, no. 4, pp. 237-250, December 1982.

64. C.L. Liu and J.W. Layland, "Scheduling Algorithms for Multiprogramming in a Hard Real-Time Environment," *JACM*, vol. 20, no. 1, pp. 46-61, 1973.

65. C.D. Locke, "Software architecture for hard real-time applications: cyclic executives vs. fixed priority executives," *Real-Time Systems*, vol. 4, no. 1, pp. 37-53, March 1992. Real-Time Syst. (Netherlands)

66. J.A. McDermid, "Assurance in High Integrity Software," in *High Integrity Software*, ed. C.T. Sennett, Pitman , 1989 .

67. Ada 9X Project Office, *Ada 9X Requirements Document*, December 1990.

68. R. Rajkumar, L. Sha, and J. P. Lehoczky, "An Experimental Investigation of Synchronisation Protocols," *Proceedings 6th IEEE Workshop on Real-Time Operating Systems and Software*, pp. 11-17, May 1989.

69. K. Ramamritham, "Allocation and Scheduling of Complex Periodic Tasks," *10th International Conference on Distributed Computing Systems*, pp. 108-115, 1990.

70. L. Sha, J.P. Lehoczky, and R. Rajkumar, "Task Scheduling in Distributed Real-time Systems," *Proceedings of IEEE Industrial Electronics Conference*, 1987.

71. L. Sha, J.B. Goodenough, and T. Ralya, *An Analytical Approach to Real-Time Software Engineering,* Software Engineering Institute Draft Report, 1988.

72. L. Sha, R. Rajkumar, and J. P. Lehoczky, "Priority Inheritance Protocols: An Approach to Real-Time Synchronisation," *IEEE Transactions on Computers*, vol.

39, no. 9, pp. 1175-1185, September 1990.

73. L. Sha and J. B. Goodenough, "Real-Time Scheduling Theory and Ada," *IEEE Computer*, April 1990.

74. S.K. Shrivastava, L. Mancini, and B. Randell, "On The Duality of Fault Tolerant Structures," in *Lecture Notes in Computer Science* , vol. 309, pp. 19 - 37, Springer-Verlag, 1987.

75. H.R. Simpson, "The Mascot Method," *Software Engineering Journal*, vol. 1, no. 3, pp. 103-120, May 1986.

76. M. Sloman and J. Kramer, *Distributed Systems and Computer Networks,* Prentice-Hall, 1987.

77. B. Sprunt, J. Lehoczky, and L. Sha, "Exploiting Unused Periodic Time For Aperiodic Service Using the Extended Priority Exchange Algorithm ," *Proceedings IEEE Real-Time Systems Symposium*, pp. 251-258, December 1988.

78. B. Sprunt, L. Sha, and J. P. Lehoczky, "Aperiodic Task Scheduling for Hard Real-Time Systems," *Real-Time Systems*, vol. 1, pp. 27-69, 1989.

79. Ada 9X Mapping/Revision Team, Intermetrics, "Ada 9X Reference Manual, Draft Version 5.7," Ada 9X Project Report, October 1994.

80. Cadre Technologies, *Teamwork,* 222 Richmond Street, Providence, RI 02903, USA, 1990.

81. K. Tindell, A. Burns, and A. Wellings, "Mode Changes in Priority Pre-emptive Scheduled Systems," *Proceedings Real Time Systems Symposium*, pp. 100-109, Phoenix, Arizona, December 1992.

82. K. Tindell, "Using Offset Information to Analyse Static Pre-emptive Scheduled Task Sets," YCS 182, Department of Computer Science, University of York, September 1992.

83. T. Vardanega, S. Ekholm, and A. Paganone, and T. Vardanega, "Experience with the Development of Hard Real-Time Embedded Ada Software," *Proceedings of the 16th IEEE International Conference on Software Engineering*, Sorrento (Italy), May 1994.

84. A.J. Wellings, "Real-time Requirements Session Summary," *Proceedings of the 4th International Workshop on Real Time Ada Issues, Ada Letters*, vol. 10, no. 9, 1990.

85. J. Xu and D.L. Parnas, "Scheduling Processes with Release Times, Deadlines, Precedence, and Exclusion Relations," *IEEE Transactions on Software Engineering*, vol. SE-16, no. 3, pp. 360-369, March 1990.

Index